FINTECH
FOR BILLIONS

Celebrating 35 Years of
Penguin Random House India

ADVANCE PRAISE FOR THE BOOK

'In *Fintech for Billions*, Bhagwan Chowdhry and Anas Ahmed have a very readable account of why India, despite the hype, has not yet brought finance to all, but also how technology, if properly applied, can make it a reality. The well-researched book has many fascinating examples of how innovations are taking place all over India to make finance accessible. This is a must-read for anyone interested in the promise of fintech for a new India, and more generally, for the world'—**Raghuram Rajan, Katherine Dusak Miller Distinguished Service Professor of Finance, University of Chicago Booth School of Business, and former governor, Reserve Bank of India (2013–2016)**

'Chowdhry and Ahmed see both the promise of Digital India and what still needs to be done. What is refreshing and helpful is they provide many constructive and pragmatic solutions'—**Nandan Nilekani, chairman and co-founder, Infosys, and founding chairman, Unique Identification Authority of India (UIDAI)**

'As they say of every revolution—for a long time, nothing happens, and then suddenly, everything happens. That's been the case with financial inclusion as well. Over the last decade, digital technologies have shown the Holy Grail of financial inclusion. In this short and engagingly written book, Chowdhry and Ahmed tell us how to make the magic of digital technologies work for meaningful financial inclusion'—**Duvvuri Subbarao, former governor, RBI (2008–13)**

'The ambition underlying *Fintech for Billions* is "India-scale"—to make Fintech work for a billion Indians. Laced with umpteen digital stories from the nook and corner of India, the book shares a simple, but profound, idea: financial inclusion can tellingly uplift lives if it imbibes the motto SHUb! An acronym for Simple, Human and Ubiquitous, SHUb, the authors develop this concept by rigorously analysing it through the economists' prism of market failures. I am incredibly excited by the auspicious promise that SHUb offers India; after all, SHUb means auspicious in Hindi. Given the genuine leapfrogging that India has already accomplished in creating digital infrastructure, read the book to share my excitement'—**Krishnamurthy Subramanian, executive director, International Monetary Fund, and 17th chief economic adviser to the Government of India**

'India lives in its villages, as Mahatma Gandhi once said. India's future therefore must encompass its citizens in rural, often unbanked areas, where they grind out a living, are employed in micro-, small- and medium-sized enterprises, and perform a fine balancing act between paying high-borrowing rates to invest in family needs with saving enough to meet unexpected shocks like a healthcare emergency. India's response to address this challenge must be low-cost, replicable or scalable, and inevitably reliant on its world-class digital finance plumbing. How can India go about making this happen over the next decade?

FinTech for Billions: Simple | Human | Ubiquitous, by Bhagwan Chowdhry and Anas Ahmed, is a must-read for anyone keen to understand the banking for a billion-plus that is underway in India, its immense promise to unleash credit and growth at the last mile, and why Indian Fintech remains the most attractive sector for foreign institutional investors to pour money into. The book is accessible and easy to read, reflects deep understanding of India's needs at the level of *every* Indian—not just the financially sophisticated, traverses with simplicity the complex path from Digital Didi to Ayushman Bharat Plus, and along this journey, presents a pathway for how every Indian can be put on the path of financial sophistication leveraging India's FinTech advantage'—**Viral Acharya, C.V. Starr Professor of Economics, Department of Finance at New York University Stern School of Business (NYU-Stern), and former deputy governor, RBI**

'The book is very insightful in terms of the strides made in the last decade of digital financial inclusion and full of recommendations on how to make it more effective. This is a must-read for all practitioners. The microfinance business has evolved significantly since the early NGO days. The business model, technology and scale of operations have changed dramatically and today microfinance is undertaken very viably by banks and NBFCs [Non-Banking Financial Companies]'—**Samit Ghosh, founder, Ujjivan Financial Services Limited**

'I've studied financial inclusion in India for twenty years, and I learnt a tremendous amount by reading this book. Accessible and highly engaging, this book fills an important gap in our understanding of financial inclusion and what works for humans.

Bhagwan Chowdhry is an eminent scholar with a deep understanding of the most technical aspects of finance. Together with Anas Ahmed, they provide the most thorough and thoughtful overview of the challenges faced by hundreds of millions of Indians struggling on the edge of financial inclusion and propose a simple, manageable framework to guide a path forward. A must-read for anyone interested in financial inclusion'—**Shawn Cole, John G. McLean Professor of Business Administration at Harvard Business School, and co-chair of J-PAL's Research, Education and Training vertical**

'India is leading the world in using technology to build financial services for the poor. This timely, well-researched and highly accessible book is filled with illuminating vignettes of how Fintech is transforming lives and includes several practical ideas to help fully realize the potential of fintech for universal and effective financial inclusion'—**Karthik Muralidharan, Tata Chancellor's Professor of Economics, University of California San Diego**

'Former Reserve Bank of India governor Y.V. Reddy coined the term "financial inclusion" while working on the draft of the annual monetary policy statement for 2005–06. Initially, Reddy had set his eyes on the "exclusion" term. But he felt that as the Central Bank had been trying to encourage commercial banks and other financial intermediaries to reach out to more people, the right term for these activities should be "financial inclusion" and not exclusion. The drive is still on.

Fintech for Billions has seen financial inclusion and exclusion through a different prism. Getting out of the classroom, the authors travelled to the hinterland in search of answers to the questions often asked and answers are vague. In the process, they have busted many myths about inclusion and exclusion. Going beyond theories, they also offer solutions to the problems, entrenched in the financial system. An easy read on what fintech can do to change the landscape at the base of the pyramid'—**Tamal Bandyopadhyay, journalist and author**

'Unleashing the power of financial technology to empower the masses, *Fintech for Billions* by Professor Bhagwan Chowdhry and Anas Ahmed is a fascinating exploration of digital inclusion. Through bold proposals like "basic accounts" for the poor and impactful roles for Banking Correspondents, this book champions a simple, human and ubiquitous approach to finance. A must-read for anyone seeking innovative and inclusive solutions to bridge the digital divide in finance'—**Raghavendra Rau, Sir Evelyn de Rothschild Professor of Finance, Cambridge Judge Business School, University of Cambridge, UK; co-editor,** *The Palgrave Handbook of Technological Finance;* **and co-founder and director, Cambridge Centre for Alternative Finance**

'The India story has been gathering pace in the global narrative over the past decade. Analysts, investment banks, think tanks and global agencies have all provided gushing salutes to the potential of India in terms of its size, workforce and geopolitical centrality. A key underpinning of this optimism is the soft infrastructure the country has put in place through an industrial-scale rollout of digital connectivity that has allowed people to link their bank accounts to a national identity number and a mobile phone number. Amongst others, this linkage is promising financial inclusion of over 500 million people at the bottom of the income pyramid. But has financial inclusion truly reached this group? This research-based effort by Chowdhry and Ahmed provides an incredible window into the reach of financial inclusion to Indians. The reality, as they reveal, is not necessarily what the popular narrative suggests. The authors illustrate the various ways in which schemes like the Pradhan Mantri Jan Dhan Yojana (PMJDY) have failed to achieve real financial inclusion for people at the bottom of the pyramid. The beauty of the book, however, lies in how the authors provide a rubric to evaluate financial inclusion schemes (SHUb—simple, human and ubiquitous) and then examine where some of these schemes fall short. My personal favourites are the chapters on the importance of a voice interface and the one on "Bank Balance Batao" which illustrates why voice matters. The book is an absolute delight to read. It is breezy, full of anecdotes and simple takeaways. If you want to get a bird's-eye view of financial inclusion in India, this book is a must-read'—**Amartya Lahiri, Royal Bank Faculty Research Professor, Vancouver School of Economics, University of British Columbia**

'This is a remarkable and timely book that puts the spotlight back on the true goals of financial inclusion and what it really means for the people that it is intended for. It deftly lays out the hits and misses of the movement in the past decade or so through compelling stories from the field. The authors have been keen observers and ideators in this space. Their case for eliminating transaction costs in banking for the poor, among others, is worth serious consideration given the paradox of growing digitization co-existing with persistently high currency in circulation'—**Bindu Ananth, chair, co-founder and managing trustee, Dvara Holdings**

'I think this is an important book. The concept of SHUb should be universally applied to government programmes that target the poor. Designing government programmes should not only focus on the efficiency of delivery but should preserve the dignity of the individual. All individuals, irrespective of their economic and financial circumstances, deserve equality in dignity'—**Mudit Kapoor, associate professor of economics, Indian Statistical Institute, New Delhi**

'*FinTech for Billions* rethinks conventional wisdom and candidly puts forth the ground-level realities of financial inclusion in India. Based on research, it delves straight into answering the pressing question: why is financial inclusion falling short?

The book offers innovative insights and recommendations that I believe are crucial for addressing the needs of the bottom of the pyramid. It is a must-read for policymakers, institutions, innovators and all those deeply invested in creating an inclusive financial system'—**Rajesh Bansal, chief executive officer, Reserve Bank Innovation Hub**

'Financial inclusion in India is lost in a maze of slogans, government coercion and computer technology. This important book takes the trouble of actually looking at poor people, understanding their life, objectives and constraints. It is rich with insights on why many things don't work, and on the first principles thinking that needs to be applied—one community at a time—for innovators to find genuine solutions'—**Ajay Shah, research professor of business, Jindal Global University and author of** *In Service of the Republic: The Art and Science of Economic Policy*

'This book has the most cogent explanation of the financial ups and downs of India's lower-income households. All done through gripping human stories . . .

Each chapter describes one or other financial service, the lack of access to these by the poor, the reasons why and most importantly, brilliant innovative solutions like a nationwide phone number 222 (BBB), on-demand Mukti loans, savings linked insurance, and Cash Ladder savings product to beat inflation'—**Vijay Mahajan, founder, BASIX Social Enterprise Group**

'In this captivating book *Fintech for Billions*, Chowdhry and Ahmed bring to light the human element essential for Fintech to truly transform over a billion lives in India. Through the SHUb concept (Simple, Human, Ubiquitous), they underscore the significance of human-centric approaches.

The authors skilfully simplify complex financial products and use the human element to meet the needs of the masses in India. The chapters on Voice and Bank Balance Batao are particularly insightful as the show us how voice can be used in Fintech for everyone. This book is a must-read for all bankers and all those seeking to harness the true potential of Fintech in the last mile'—**J. Venkatramu, chief executive officer, India Post Payments Bank**

'*Fintech for Billions* suggests practical solutions to ensure last mile financial inclusion and help every Indian easily access financial products and services that can transform their lives'—**Shilpa Kumar, partner, Omidyar Network India and Former managing director and chief executive officer, ICICI Securities**

'India has seen a decade or more of investment in digital infrastructure. The authors rightly shine a light on the human capital limitations of using such infrastructure. With basic economic reasoning, they make the case for practical, low-cost interventions—think Microangels and Digital Didis—that can improve financial inclusion for "billions"'—**Tarun Khanna, Jorge Paulo Lemann Professor at the Harvard Business School and author of** *Billions of Entrepreneurs: How China and India Are Reshaping Their Futures and Yours*

FINTECH
FOR BILLIONS

SIMPLE ■HUMAN ■UBIQUITOUS

Bhagwan Chowdhry
and Anas Ahmed

PENGUIN
VIKING
An imprint of Penguin Random House

PENGUIN VIKING

USA | Canada | UK | Ireland | Australia
New Zealand | India | South Africa | China | Singapore

Penguin Viking is part of the Penguin Random House group of companies
whose addresses can be found at global.penguinrandomhouse.com

Published by Penguin Random House India Pvt. Ltd
4th Floor, Capital Tower 1, MG Road,
Gurugram 122 002, Haryana, India

First published in Viking by Penguin Random House India 2023

10 9 8 7 6 5 4 3 2

ISBN 9780670096213

Typeset in Adobe Caslon Pro by MAP Systems, Bengaluru, India
Printed at Replika Press Pvt. Ltd, India

www.penguin.co.in

The authors' revenue generated from this book will be donated to an organization engaged in spreading financial literacy, consumer awareness and advocacy.

The views and opinions expressed in this book do not necessarily reflect the views and opinions of any affiliated organizations the authors are associated with or have been connected to in the past.

Contents

Introduction

We'll just say it as it is: don't believe everything you read and watch in the media about financial inclusion. All the newspaper coverage, online posts, and web and journal articles will have us believe that financial inclusion has arrived for the Bottom of the Pyramid (BOP) population.

It has not.

Well, not yet.

The number of bank accounts opened under the Pradhan Mantri Jan Dhan Yojana (PMJDY) is around 49 crore (490 million) and the aggregate account balance in them is almost Rs 2 lakh crore ($24 billion). This may look good per se.[1] Still, they do not reflect the realities of dormant accounts (almost one out of every five accounts remains dormant),[2] coercively

---

[1] 'Pradhan Mantri Jan-Dhan Yojana | Department of Financial Services | Ministry of Finance', n.d., https://pmjdy.gov.in/account, accessed on July 21, 2023.

Sridhar, G. Naga, 'Total Balance in Jan Dhan Accounts Sees Record Spurt in FY23', *The Hindu Business Line*, April 9, 2023, https://www.thehindubusinessline.com/money-and-banking/total-balance-in-jan-dhan-accounts-sees-record-spurt-in-fy23/article66717113.ece, accessed on July 21, 2023.

[2] 'Pradhan Mantri Jan Dhan Yojana (PMJDY)—National Mission for Financial Inclusion, Completes Eight Years of Successful Implementation', n.d., https://www.pib.gov.in/PressReleasePage.aspx?PRID=1854909, accessed on July 21, 2023.

opened accounts that are rarely serviced,[3] and the reluctance of bank officials to open such accounts after a publicized drive. Publicity campaigns and sometimes even media coverage tend to overlook, even hide, the realities on the ground.

The promise of 'five minutes to open a bank account' makes for an excellent copy, but in reality, it can take several days and several visits to the bank, and even then, it may not happen.

When there is an account-opening drive organized by the banks, the accounts are opened. In fact, they are opened by millions of people because the directive has come straight from New Delhi. Banks are reluctant to open and service these low-return accounts but must adhere to the government's mandate. In fact, during our research, we encountered several public and private sector banks unwilling to open new Jan Dhan accounts when approached. We then discussed this issue with the branch officials and senior management of the banks and were told they prefer not to open these accounts as they make very little money for the banks.[4]

Data Is Misleading

Laxmi is a mother of two young children. Her husband died in an accident a few years ago. She supports herself and her family

---

[3] Staff, Scroll, 'Not Just the One-Rupee Trick: How Banks Are Hiding Jan Dhan's Zero-Balance Accounts', Scroll.in, September 14, 2016, https://scroll.in/article/816407/not-just-the-one-rupee-trick-how-banks-are-hiding-jan-dhans-zero-balance-accounts, accessed on July 21, 2023.

Sharma, M., Giri, A. and Chadha, S., 'Pradhan Mantri Jan Dhan Yojana Wave III Assessment', n.d., Microsave, https://www.microsave.net/files/pdf/PMJDY_Wave_III_Assessment_MicroSave.pdf, accessed on July 21, 2023.

[4] Moneycontrol.com, 'Banks Struggle to Make Jan-Dhan Accounts Cost-Effective and Viable, Eight Years On', Moneycontrol, n.d., https://www.moneycontrol.com/news/business/eight-years-on-banks-struggle-to-make-jan-dhan-accounts-cost-effective-viable-8621681.html, accessed on July 21, 2023.

by working as a cook and a housekeeper for two middle-class families in Hyderabad and earns Rs 12,000 (approximately $146) a month. Laxmi studied until the third grade but can barely read Hindi and has never felt the need to communicate by writing. She uses a feature mobile phone to communicate via voice calls with the members of the families who have employed her when she is late or not able to go to work for personal, family or health-related reasons.

Laxmi has an Aadhaar card—she was able to enrol several years ago. She, as well as her husband, were also persuaded to open a basic bank account under the PMJDY in 2017 when there was a big drive by the Government of India to enrol millions of people who did not have bank accounts (her husband was alive then).

When statistics are collected, Laxmi is one of many people at the BOP who are counted as financially included. She has a bank account, a mobile phone and an official identity. She is employed in the informal workforce and can make a meagre living to meet the necessities of food and shelter for herself and her family, though barely so.

Despite where Laxmi shows up in official statistics, we think she is not financially included in any meaningful sense.

Not yet.

Cash Is Still King

Though Laxmi has a bank account, she hasn't ever deposited money in it. She receives her salary from her employers in cash, which she prefers. She keeps the cash at home and spends it throughout the month, primarily to buy food and a few other essential household items. So, the cash balance begins at about Rs 12,000 when she receives her salary and goes to near zero by the end of the month. One might estimate that the average cash balance at Laxmi's home is close to Rs 4000–5000 (approximately $48–$60)—less than half of her salary because expenditures are concentrated in the early part of the month. Had she kept her cash balances in an interest-earning deposit account, she could have earned an interest of Rs 300–400 (approximately $4–$5) over the year. She earned zero—an indication that she is financially excluded.

It Hurts When She Needs to Borrow. Suddenly.

The amount foregone in interest earnings may appear to be relatively small—we are talking about less than half a percentage point per month. That is not where financial exclusion hurts people like Laxmi the most. Having a savings cushion for an unexpected cash need can be a lifesaver because Laxmi's access to affordable lending options is severely limited. She would be forced to borrow at an exorbitant rate in the informal market if such an emergency arose and she had no savings of her own.

When her husband died in 2017, she did not have enough cash savings to pay for his funeral. Her extended family helped a bit, but she still needed an additional Rs 5000 (approximately $60). A *kirana* (local grocery store) owner, where she usually buys food and household supplies, introduced her to a local moneylender, who gave her the Rs 5000 she needed with the stipulation that the total principal loan amount she was to return would be Rs 5200 (approximately $62)—implying a 4 per ent initial fee. In addition, she will have to pay interest at a monthly rate of 3 per cent on any outstanding balance. She figured that if she could manage to save Rs 200–300 (approximately $2–$3) per month, she would be able to repay the loan, interest plus principal, in about a year. She borrowed from the moneylender at an effective annual rate of nearly 50 per cent, taking into account both the initial 4 per cent fee and the recurring 3 per cent monthly interest, more than double the borrowing rate businessmen were paying—about 24 per cent a year—in the informal lending market for small enterprises. Another indication that Laxmi is not financially included.

Her husband's death was not the only time Laxmi had to borrow money from the local moneylender. Her three-year-old child fell ill with pneumonia last year. At first, when Laxmi noticed the symptoms—fever with chills—she took two days off from work to take care of him with home remedies. Around four days later, when the condition did not improve and, in fact, became worse, she panicked and took him to a neighbourhood doctor, who reprimanded her for not bringing her son for a check-up sooner. He advised her to get her son admitted to a hospital that was several kilometres away from her home. She did. Her son received proper treatment, but it cost Laxmi nearly Rs 2500 (approximately $30), which she ended up borrowing from the same moneylender she had borrowed from before. Laxmi was trying to save the cost associated with the visit to the doctor and the cost of medicines she would have had to buy—perhaps a couple of hundred rupees. She ended up paying four times as much, not to speak of several days of absences at her employers, who were kind but were greatly annoyed at the inconvenience they faced because of her multiple-day absence. If Laxmi had access to inexpensive borrowing or affordable health insurance, she might have been more inclined to get medical treatment for her son right after noticing his symptoms. The lack of financial inclusion affects Laxmi in many insidious ways.

Financial Products Are Too Complicated

When the PMJDY was introduced, one of the incentives to open an account was that account holders would be enrolled, free of cost, in a life insurance coverage of Rs 1 lakh (approximately $1219) in case of death in an accident. When Laxmi's husband died, both he and Laxmi had active Jan Dhan accounts with small balances that they had initially deposited, but Laxmi didn't receive a life insurance payout. It turns out that they never got enrolled in the free life insurance coverage because no one told Laxmi about it and her husband because they never used their RuPay debit cards that came with their accounts. However, she vaguely recalls her late husband mentioning that the eligibility criteria seemed complicated. So, although they were eligible, neither was actually enrolled—another indication of financial exclusion.[5]

It Also Hurts When She's Forced to Buy Small Quantities

Laxmi buys vegetables from the local market daily and other food staples such as rice, flour, oil and spices in quantities that last her up to a week from the kirana owned by a friendly and helpful man she has come to know well and trust—it was he who had introduced her to the moneylender when she needed cash in an emergency. The kirana owner has begun to worry about his business as e-commerce is starting to make inroads, with more and more people choosing to buy groceries online, which turns out to be cheaper—he has begun to source his inventory from online merchants. Fortunately, his clientele is people like Laxmi who buy small quantities at a time with a higher per-unit price than she could obtain by buying larger, more optimal quantities. This very intermediation, in which the kirana owner obtains supplies in large quantities and sells them at a higher per-unit price to his clients, is a big part of his earnings. The kirana owner makes his living because customers

[5] Nair, Sobhana K., 'Only Half of Pradhan Mantri Jan Dhan Yojana Insurance Claims Settled in Two Years', *The Hindu*, April 27, 2023, https://www.thehindu.com/news/national/only-half-of-pradhan-mantri-jan-dhan-yojana-insurance-claims-settled-in-two-years/article66777561.ece, accessed on July 21, 2023.

like Laxmi are not able to borrow money at affordable rates to be able to buy larger quantities of food supplies such as oil, rice and flour and store them at home to last a month or two at a time rather than just a few days, paying nearly 25 per cent more on average compared to what they would pay if they were to buy larger, more optimal quantities. That amounts to a monthly cost of approximately Rs 1000–1500 (approximately $13–$18) a month. For Laxmi, this is a substantial cost for not being financially included.

Why Pay Insurance Premiums? God Is There, and So Is Family

In the last decade, we have seen the introduction of many health insurance plans that cover large unexpected medical costs associated with secondary and tertiary care. Laxmi and others like her do not enrol for these plans because they require an upfront monthly cost in the form of insurance premiums, and the potential benefits seem distant, uncertain and often not even visible—Laxmi has never heard of her friends or family receiving payments from any insurance company. For people like Laxmi, it is only their immediate and sometimes extended family that comes to their aid if they are lucky. And indeed, Laxmi considers herself lucky that she managed to get her son treated on time, even if it cost her money.

Why Is Financial Inclusion Not Working?

Why has a much-feted and widely publicized drive to include the under-represented failed to trickle all the way down to its intended recipients? Why is there an overwhelmingly large number of people in the BOP group who do not yet have access to or entry into, the financial system that people like you and us use and benefit from every day? Why have our well-intentioned plans not worked fast enough?

Throughout this book, we will explore why many of the existing solutions have faltered and fumbled along their path to inclusion. We will reveal, through what we now call the **SHUb Test** (*shub* also means auspicious in Hindi) that each of these solutions has failed. Through the course of our research, which was designed to ascertain the causes of financial exclusion and to propose alternate solutions that would aid in inclusion, we found that any solution, tool or technology that was deployed for the BOP group needed to pass through three filters:

It must be **Simple**.

It must be **Human**.

And it must be **Ubiquitous**.

In short, it must be **SHUb**.

Simple: The first filter is that of simplicity. Banking and financial products need to be simple in order for financial inclusion to be successful because simplicity allows more people to understand and use these products. When solutions are complex or confusing, it can be difficult for people to understand how they work, what they are used for and how to access them. This can lead to a lack of trust in these products, which can prevent people from using them.

Simplicity is essential for financial inclusion because it allows people who may not have a lot of education or financial literacy to understand and use these products. When products are simple, it is easier for people to make informed decisions about their finances, and it is more likely that they will use these products to improve their financial well-being.

In addition to helping people understand and use these products, their simplicity also makes it easier for financial institutions to provide them to a wider range of customers.

When products are simple, it is easier for financial institutions to educate their clients and staff about how to use them, and it is easier for them to manage and administer these products. This can help financial institutions reach more people and provide better services to their customers.

Human: The second filter is that of the human touch. Financial products must have a human touch if we want them to be inclusive because this is what helps build trust and improve the experience of the BOP group. When people have access to banking and financial products that are personalized and provide a human connection, it can help them feel more comfortable and confident in using these products. This is particularly important for people who may not have a lot of experience with financial products or who may be hesitant to use them.

Human touch can also help provide support and guidance to customers who may need assistance with using these products. This can include providing advice on how to manage their finances, helping customers understand how to use different products and addressing any questions or concerns they may have. When people feel they are being supported and they have someone to turn to for help, they are more likely to use these products and trust the financial institutions that provide them.

Ubiquitous: The final filter is ubiquity. Financial solutions need to be ubiquitous in order to ensure people have access to these products regardless of where they live or what their circumstances are. Ubiquity refers to the availability of something everywhere or the fact that it is present in all places at all times. When banking and financial products are ubiquitous, it means that they are available to everyone and that people can access them easily and conveniently.

This is particularly important for financial inclusion because it ensures people who may not have access to traditional financial institutions or who may live in remote or underserved areas can still use these products. It can also help to reduce the costs associated with accessing these products, as people do not have to travel long distances or incur other expenses to use them.

For financial products to be ubiquitous, they must be available through a variety of channels, including physical branches, mobile banking apps and online platforms. This allows people to access these products conveniently. Also, it helps ensure these products are accessible to everyone, regardless of their location or circumstances.

If a product or service fails to pass through even one of these three filters, it is unlikely to have a significant impact on financial inclusion.

Here are examples of how existing solutions fail on one, two or all three of these.

What Is SHUb?

Cash is accepted everywhere. Cash is simple and intuitive. You can't accidentally hand over cash to the wrong person. You can't accidentally give the wrong amount either, because almost all of us usually count several times before giving cash to someone. Cash passes all three tests: it's Simple, Human and Ubiquitous. It's SHUb.

Despite excelling on all three parameters, cash is still limiting. It is local and can only travel at the speed of the person who has the cash. Cash does not allow for e-commerce. If one relies entirely on cash, they are excluded from all the wonderful things promised by the digital revolution.

What Is not SHUb?

Electronic payments are not SHUb. Albeit simple, they lack human touch and are not ubiquitous. Sometimes, bank staff are rude to poor and uneducated people, so they fail the Human test. Business Correspondents (BCs), who exemplify the human touch, are not ubiquitous. Contrarily, moneylenders are ubiquitous and simple, but on account of their steep interest rates, they cannot be considered humane.

The PMJDY has an overdraft facility, but most people don't know about it. Even for those who do know about it, the conditions are onerous and the paperwork daunting. They fail on possibly all three filters of SHUb. Likewise, formal lending products are neither simple, human nor ubiquitous.

Through the course of this book, we have explored some existing solutions as well as recommend a few. In the chapters that follow, we will examine what financial inclusion means and identify the degrees to which one could be financially excluded.

Our research has taken us to all corners of India, from towns in Rajasthan to villages in Goa, from hamlets in Odisha to districts in Himachal Pradesh and Telangana. We've met with individuals from the bottom of the pyramid, frontline employees at banks' branches and service points, mangers and heads of divisions with various financial service providers, senior executives, regulators and the government.

In addition to our qualitative interviews and discussions with the above-mentioned stakeholders, we have also relied on secondary research as well as quantitative surveys and administrative data originating from the pilot studies of a few solutions we have been conducting in partnership with various financial service providers to develop our recommendations.

Throughout the book we attempt to provide a human account backed by research and interviews, as it is our endeavour to dispel myths around the inclusion figures as well as to reinforce that inclusion need not be a myth at all—beyond the grasp of policymakers, economists and financial institutions. Inclusion can be a very attainable and reachable goal.

Chapter One

Voice

Voice plays a very significant role in human interactions. Our voices are not just sounds produced in the larynx; they are also very integral to our personalities. The specific tone and timbre of our voices give them a distinct identity, much like our fingerprints. In fact, on several occasions, investigative agencies have relied upon the scientific breakdown of voice indicators to narrow down suspects. Sophisticated software programmes in forensic laboratories can match voice samples with near certainty, and this technology is no longer limited to military and government use but is seeing widespread commercial adoption globally by organizations to authenticate and protect their customers from fraud.

There are other ways in which voices act as identity markers. There are para-linguistic and non-linguistic indicators as well. For instance, by simply listening to a voice, you can establish the gender and approximate age of the speaker. Accents, which are inextricably tied to our voices, can help in placing an individual at an approximate location on the globe with a fair amount of accuracy.

Voices have such a distinctive quality about them that many of us have memories related to them. We can clearly recall the stories that were narrated to us in the distinct voices of our family members, parents, or even beloved grandparents who may have passed away. Sometimes, even famous voices have recall: in India, for example, singers such as Kishore Kumar, Lata Mangeshkar, Asha Bhosle and Mohammed Rafi enjoy legendary status because of their inimitable voices. In Indian radio, Ameen Sayani has enthralled audiences across generations. In the field of sports, the voice of the much-loved cricketer Sachin Tendulkar may not pack a punch like that of artistes known for their voices, but it is still instantly recognizable and eminently trustworthy.

After all, we hear voices much before we 'see' anything. There is evidence from studies conducted at the University of Helsinki and the University of South Carolina to suggest that children in the womb can not only hear voices but also distinguish between different voices as early as thirty-two weeks of gestation. Furthermore, they found that foetuses demonstrated a preference for human voices over computer-generated speech while in the womb.[1]

Radios may not be as popular as they were decades ago, but podcasts have picked up. So many of us choose to turn on an application to listen to soothing sounds before we go to bed, or even listen to a podcast in the sonorous voices of our favourite artists. Such is the effect that voices can have on humans.

A case in point is Disney movies. Disney has largely been making animated movies, but these have done as much business as non-animated features with real people acting in them. It can be argued that it is the voices that make the animated films appear almost real and give them that lifelike quality that Disney is known for. That, and, of course, the superior animation. It's believed that the actors who lend their voices to animated movies are paid on par with their other acting roles. Children find an emotional connection to voices rather than real-life visuals, which Disney movies have successfully harnessed for decades.

Animated movies are not the only places we encounter 'voices' without the people they belong to. Many of us might

[1] Lehtonen, L., Leminen, A., and Peltomaa, R., 'Fetal movements in response to the maternal voice', *Acta Obstetricia et Gynecologica Scandinavica*, 88(4) (2009): 468–473.

Bower, S.S., Schiefelbusch, D.L., Warren-Lemasters, N., Stiefer, G., and Schnelle, J., 'Preference for maternal and other human voices by human late-gestation fetuses', *Early Human Development*, 28 (1992): 99–109.

instantly recall the unique voice that we've heard in the recorded IVR of our cellular phone provider or the very distinctive voices that announce station names on the Metro lines.

In early 2021, when the Government of India launched the COVID-19 vaccination drive, it was nothing short of a mammoth task. The goal of vaccinating the entire population was tremendous, to say the least. In times like these, when superstitions were rife because of inadequate public knowledge, appealing to people's better sense alone would have been woefully inadequate. The government needed something more than just warnings to get the masses to take heed. And as importantly, the large population needed to be sensitized about controlling the spread of the virus by adopting hygienic practices.

It was at this time that Amitabh Bachchan, arguably India's most iconic film star, was roped in to deliver a public service announcement in his distinguished voice. It was perhaps as much for Bachchan's unmatched voice as it was for his personal brand. He is no stranger to public service messages. Over his long and illustrious career, he has advised, cautioned, encouraged and counselled people variously on the importance of donating eyes to the absolute necessity of administering polio drops.

A voice is associated with humanness, personal connectedness and intimacy. Certain voices may motivate, induce trust or even create a sense of security. Amitabh Bachchan embodies all of these, and perhaps that is the reason behind the success of his voice.

This is perhaps why we find that a lot of service providers who are replacing manual processes with digital solutions continue to rely on some form of voice interaction to help customers feel connected to a 'person'. Alexa the smart speaker, for instance, is nothing but a computer program. And yet, the 'intelligence' with which she listens to questions and the wit with which she responds to them keeps users engaged. She sounds so 'human' that more than one of us has tried having entire conversations with the voice of Alexa. Several elderly people often say 'thank you' to Alexa for helping them find answers.

This urge to engage with 'Alexa', which most of us know to be an AI-generated voice tool, is not accidental. As humans, we are programmed (a different kind of programming from Alexa's, of course) to communicate. When we encounter a voice, there is a need to talk to that voice. And our voices are critical inputs in the entire back and forth with Alexa. There's also the question of ease—we don't have to type out commands and instructions anymore; we can simply 'tell' Alexa what we need. While this was no doubt built for ease, the possibilities for using a technology such as this for groups with limited literacy are immense. Not only can the user feel that they're speaking with a 'human', but they also feel more a part of the process.

In essence, because of the way humans are wired for social connections, voice is a tool that can engage people even when there is no real human presence. Voices are unique, and they are calmer and more reassuring than machines. We intuitively trust voices more than we do machines, processes and automated replies.

Why Voice?

There is another, more practical use of voice in the service sector, particularly in the banking arena. When dealing with customers who have low or marginal literacy, a voice comes in very handy. Even a customer who's unable to read can listen to oral instructions and respond to them on IVR calls, self-help machines, or even ATM kiosks that allow for a voice function. If they do not guide the customer entirely through, say, a withdrawal process from the ATM, they at least support the same transaction.

Bhishma is a vegetable vendor in a small town in Maharashtra. He runs his shop on the pavement every day for a few hours early in the morning and another few in the evening. Those are the only hours when he can sell without the fear of a policeman making him shut down his efforts. He doesn't have a vendor's licence yet. Since it was becoming difficult for him to manage his expenses and send some money back home due to the little money that he can make in these few hours, he finally decided to take up day-time employment as a helper at a local wholesale store to supplement his earnings. He does have a bank account, though. Every month, he gets help from one of the six men he shares his room with to send money home.

Even though he has been living in the city for a few years now, he hasn't quite gotten used to the technology of his phone yet. He is also intimidated by ATMs. He usually asks for help, but on a few occasions, he's had to transact by himself. The last time he went to the ATM, he wanted to withdraw Rs 1000. The queue behind him was daunting, as was the security guard at the ATM kiosk, who was keeping a watchful eye on him lest his inability to transact confidently delay the rest of the people behind him. Bhishma got flustered and punched in what he thought was an amount for Rs 1000. But when the money came out of the machine, he got even more flustered to see so many currency notes. He had accidentally withdrawn all his money. This was the money he was planning to send home later that same week. He tried asking the security guard a few questions, but the guard was abrupt with his responses. He told Bhishma that the only way the machine would give him Rs 10,000 (approximately $122) was if he had asked for Rs 10,000.

What is likely to have happened is that Bhishma would have accidentally pressed '0' one time too many. This is not a big problem, you'd think. A simple deposit transaction can fix the accident. However, what is a minor inconvenience for you and us, dear readers, is a giant task for someone like Bhishma. He will now have to visit a bank branch to make the deposit, which is a few kilometres away. Since it's the first week of the month, he could end up spending over an hour waiting in line, and hence he would need to take some time off from his new employment at the wholesale shop, which means he will be at the mercy of his supervisor to give him the time off on a weekday. God forbid if it takes longer than a few days. Then his family waiting in the village will start to get impatient.

You might be thinking that surely the ATM would ask for confirmation before giving out the cash. In fact, in Bhishma's case, there was a pop-up on the screen asking that very question. However, semi-literate people like Bhishma would not be able to grasp the written text on the screen in a single glance, especially in a flustered state. Even a literate person might make the same mistake as they so often do. Perhaps it's because we choose what we see and read, but a voice just cannot be ignored.

Written communication is usually viewed as a push factor. When a customer is faced with text in any form, they have a choice as to whether they want to read it or not. It could be text on the screen of an ATM, a text message on WhatsApp or even an SMS. On most occasions, recipients may choose to read the message later or ignore it entirely. Listening, on the contrary, is a pull factor. If there's a voice that the customer can hear, their mind responds to it. The reaction to a voice is almost reflexive. In Metro trains, for example, there is a voice that warns users when it announces before every station, 'Doors will open on your right' (or left, as the case may be). A similar text warning flashing just above the door may not be as effective, leading to inevitable crowding and confusion when passengers get off and on simultaneously.

When dealing with customers who have low or marginal literacy, a voice comes in very handy. Even a customer who's unable to read can listen to oral instructions and respond to them on IVR calls, self-help machines, or even ATM kiosks that allow for a voice function. If they do not guide the customer entirely through a withdrawal process from the ATM, they at least support the same transaction.

Accuracy is much higher as well when voice instructions are involved, which could impact the overall experience of the customer. It is true that a lot of AI-generated voices sound mechanical, monotonous or even impersonal. But AI is making strides in incorporating tonality, emotions and feelings to ensure that even a computer-generated voice sounds human enough to make the customer feel a sense of warmth and to make them feel welcome and connected. The most recent advances in AI are even able to clone the voices of specific humans with just ten minutes of voice samples for training.

In fact, adding voice as a tool to make the transaction error-proof is an example of *Poka Yoke*, which is a Japanese term that refers to techniques used to prevent mistakes or errors in manufacturing and other processes. The term itself can be translated as 'mistake-proofing' or 'error-proofing'.

One of Poka Yoke's key principles is to design processes in such a way that errors are either impossible or easily detectable and correctable. This can be achieved through the use of physical or mechanical devices, such as jigs or fixtures, that ensure components are assembled correctly or through the use of visual aids, such as colour-coding or special markings, to guide people through a process.

Existing Use of Voice

In the early days of digital payments, shopkeepers and small traders would track receipts based on text messages received from banks. Invariably, this would take some time. Networks were often slow or patchy, and there was no way of ensuring that the money had actually been credited to the seller's account. The systems were still new, as was the method itself, and there was considerable doubt in the minds of some users as to whether the money would come in. Even if the customer showed the phone screen with a debit, it just wasn't enough to convince the shopkeeper—what if the money did not come into their account, or what if the transaction got reversed before it could come through? Scenes of customers making payments and then impatiently waiting at the counter until text messages confirmed the transfer was not uncommon.

This is also how Mushtaq Ali's little shop off the highway operated in those early days. Being next to a petrol pump meant a steady flow of customers. But the heavy footfall also meant delays in checking for payment receipts. Being on the highway, Mushtaq didn't really have regular customers. Most of them would just stop on the way. The concept of *hisaab* (keeping account) wouldn't work with one-time customers. It was no use insisting on cash either. In the wake of the pandemic, people were terrified of handling currency notes and coins. In fact, the pandemic was the very impetus behind the booming digital transactions. Mushtaq had recently procured the QR code to accept Unified Payment Interfaces (UPI) payments for his shop.

What is UPI?

Unified Payments Interface (UPI) is a real-time payment system widely used in India. It allows users to link their bank accounts to a mobile application and enables seamless and instant fund transfers. With UPI, users can initiate transactions using a mobile number or by scanning a QR code. It supports person-to-person transfers, merchant payments, bill payments, and more. UPI is interoperable, allowing users to transact across different banks and platforms.

But the delay in the text messages annoyed customers who had to wait until the confirmation message was received. Queues sometimes got rowdy. Mushtaq sometimes lost customers who couldn't even come to the front because a previous customer was still haggling about how the money had gone from his account.

After the first few weeks, Mushtaq sent for his teenage nephew from his village. Farhad's only job was to sit slightly away from the shop, where the network was better, and monitor Mushtaq's phone to track the text messages that confirmed transactions. This allowed Mushtaq to focus on the customers. Besides, the ones who haggled over transaction alerts were no longer doing it at the expense of waiting customers.

Mushtaq then installed the Paytm Sound Box after hearing about it from his friend. The Paytm Sound Box is a QR code display with a voice-enabled speaker that can announce, in over ten Indian languages, the amount credited to one's account as soon as a customer makes a transfer. With improved cellular service technology, aided by the voice mechanism, the receipts that Farhad tracked earlier are now recorded immediately and announced for the benefit of both parties. Farhad is now free from his responsibility at his uncle's shop. However, he didn't want to return to his village, so Mushtaq enrolled him in the school nearby. In the evenings, Farhad sits next to his uncle in the shop and learns the trade while completing his schoolwork.

In fact, voice-enabled speakers are now so pervasive that it's not uncommon to see regular customers at small shops and tea stalls wordlessly ask for their usual items and quickly use their smartphones to scan the app and pay for the purchase—carrying out the entire transaction in complete silence. Driven by the efficiency of the voice-enabled machine, small tea stalls too can be seen accepting digital payments for as little as Rs 5–6 (approximately $0.061) at a time.

There is another benefit of such voice-enabled systems, which was best explained by Preeti, a banana seller, during a conversation with Prime Minister Modi, '*Achaa hai na?* . . . *Saamne waala paagal nahi banaa sakta humeh* [Isn't this better? . . . I can't be fooled now].'

Preeti was referring to the scams that occurred during the early days of the UPI apps. Just before making their payments, some customers would distract the shopkeepers and surreptitiously exchange the small QR code displays in the front for one of their own. In effect, they would be making a payment to themselves. Shopkeepers, who are usually in a hurry to serve the next customer, would only glance at the screen presented by the customer to confirm that a transaction for the exact amount of the purchase had been completed. Unknown to them, they were not receiving any of the money. Now that shopkeepers can use voice-enabled devices, scams like these have been stopped in their tracks.

It's not just payment systems that are using voice now. Several other digital-only solutions have started to incorporate voice into their solutions. Raghu works as a delivery agent with the popular food delivery app Swiggy. He rides his motorbike all over Indore, and with each passing day, he gets to know neighbourhoods a little better. Since he's recently moved to the city, he's not very familiar with the city to begin with. Google Maps are helpful in general but not always accurate. Sometimes customers, too, provide incomplete addresses. Areas with dense clusters of homes and irregular numbering systems are a nightmare for Raghu.

On more than one occasion during each shift, Raghu has to call the customer and take directions from them while he simultaneously rides his bike in heavy traffic. Sometimes, he has to stop and survey the area. On a few occasions, searching for the address costs him a few extra minutes as well as the waiting customer's ire.

However, things have become much easier for him since Swiggy added the voice feature that enables customers to leave recorded voice messages with detailed directions to their homes. Raghu does not have to stop and make calls anymore. He can listen to directions on the go. It makes him deliver sooner. He is less distracted while riding, and most important of all, his customers are happy and give him high ratings.

WhatsApp is another tool that allows so many more people with a range of literacy skills to communicate easily. Before WhatsApp, there was either text messaging or regular phone calls. Call costs were exorbitant, especially for international ones. Texts were relatively more inexpensive but were limited to the literate as well as the digitally savvy. For the less literate or 'number-literate' population, WhatsApp came in as a saviour. It not only offers the option of sending voice messages instead of text messages, but even video calls are simple to make and receive thanks to easier internet access and penetration.

Parteet Rani is an elderly woman who lives on the outskirts of Gurdaspur in Punjab. The early onset of arthritis pushed her into early retirement from the family business of sugar production. She has spent most of the last decade and a half on a cot, being intermittently tended to by family members. She never learned to read or write, so spending time reading was not an option. When one of her sons bought her a phone, she quickly memorized the keys that were required to be pressed for sending voice notes on WhatsApp. Like many of her neighbours, she too had family abroad. Her daughter was settled in Canada, and Parteet missed spending time with her three grandchildren.

In the early days of the cell phone, they used to make some calls, but they quickly stopped as the phone bills were exorbitant in the first few months. Ever since the voice messaging tool has become accessible to her, she spends a lot of time recording messages, songs and even recipes and sending them out to not just her daughter but several other children from the village who have moved abroad. She feels connected to them despite the different time zones, and she looks forward to them waking up and sending her replies. She makes video calls, too, with some help, but she prefers voice notes. She feels more in command, she can marshal her thoughts more easily, and most importantly, she can send voice notes at her own pace when she has the time.

It is Parteet's experience that even when she cannot meet and talk face-to-face with her family and friends, just listening to their voices and sending them hers makes up for it. Because of the frequency, she feels 'closer' to them—something that a letter exchange is completely incapable of doing.

Payments Over Voice and UPI 123

Smartphones are expensive, and the poor can only afford feature phones. An entry-level model of a feature phone costs around Rs 1000. Besides, in remote parts of the country, where power cuts are frequent, feature phones have another advantage with longer battery lives than app-filled smartphones.

As one moves from semi-urban landscapes to more and more rural areas with limited means to buy, limited internet connectivity and limited power availability, feature phones become more common. In fact, these smaller villages and remote areas use more voice-enabled functions, such as actual calls. They do not use or need most of the features that come with a smartphone. There's another advantage that Voice has— it can be multilingual. A voice-based solution means that the beneficiary does not need to learn how to use an app or even be able to read and write.

According to the NPCI (National Payments Corporation of India) at the launch of UPI 123, there are approximately 400 million (40 crore) feature phone users in India.[2] In other words, these 400 million (40 crore) users do not have internet-enabled smartphones like the rest of us. However, this very large group of people, who may also coincide with the Bottom of the Pyramid (BOP) group, need not be excluded from the digital financial solutions that the rest of us have access to. Keeping the inclusion of this very group in mind, the NPCI launched UPI 123PAY, which is a payment system that enables feature phone users to access and utilize the UPI service in a safe and secure manner.

With UPI 123PAY, feature phone users now can conduct a range of financial transactions, such as money transfers and bill

---

[2] This number has reduced since the launch.

payments, in multiple languages using one of the many voice-based payment service providers. These voice-based technology alternatives allow feature phone users to conveniently access the UPI payment services and carry out transactions without a smartphone or internet.

Hemant Desai is a coconut water seller in a quiet little village outside Satara, Maharashtra. He has a smartphone and can accept payments through Google Pay, Paytm, etc. However, he often has customers who do not have smartphones to scan the code and make payments. Hemant has opted for a slew of digital options to ensure that receiving payments is never a problem for him, no matter what kind of phone his customers have. Of course, when he can, he takes cash, but given how scarce ATMs are, his customers sometimes choose to retain the cash and pay him digitally. Customers of Hemant who do not have a smartphone or find it difficult to use an app to make UPI payments can now choose to register on UPI 123 and make UPI payments to him by dialing the UPI 123 number on their mobile and using the IVR system to enter the payment details like amount, receiver's number and their pin to complete the payment.

Voice Infrastructure

Samita has recently moved back from the US and is settled in Gurugram (Gurgaon). Though she has a consulting practice here in India now and banks with two Indian banks for ease of business, she still has active bank accounts in the US and often transacts through them as well.

Samita has noticed quite a few differences in the way banks operate in the US and here in India. One of them is the convenience of voice recognition. Each time she calls her bank in the US, she does not have to provide detailed authentication by punching in difficult-to-recall numbers into the IVR. The advanced voice-recognition technology automatically verifies her and makes her transactions quicker and more efficient.

This is contrary to the experience of her house-help Shanti, whom Samita has overheard on several occasions trying to make a call to the bank. Shanti is linguistically challenged to begin with. Her Hindi is just passable—even Samita struggles to understand her and give her instructions. Shanti speaks fluent Bangla, though, as she comes from what she calls a 'border village' in West Bengal. More than once, Samita has overheard Shanti try unsuccessfully to navigate the IVR. Once, she even tried to speak with the recorded voice—such is her inability to comprehend banking processes. Shanti has tried on numerous occasions to punch in numbers, but she ends up getting disconnected.

Once, Samita even tried to help her by taking the phone and trying to navigate the IVR. But even Samita cannot get the information out of Shanti—asking her for her account number and card details only gets her blank stares.

Samita cannot help but wonder how much easier it would be if the bank had voice recognition, the same way hers does, even though she is fully capable of operating her account by taking an extra minute to enter details into an IVR for verification. While the voice recognition technology makes Samita more efficient by saving her a few minutes, for someone like Shanti, it could mean the difference between being excluded or included in the financial system.

One may wonder how difficult it may be for financial service providers to build solutions for Shanti and people at the BOP, as this would require the entire voice-based infrastructure to be built around Indian languages, making it a costly affair. However, there are several organizations, including government departments, research institutions and private companies that continue to build and develop the voice infrastructure around the Indian languages for further application by businesses and individuals, and we will explain some of them in the following section.

The Navana Experience

Navana Tech has worked on several solutions with financial service providers to make interactions with technology simpler and easier for those with limited digital literacy. Their technology enables last-mile access to digital services for end users through the use of voice-based technologies and products. It uses AI to build text-free, image-based and voice-assisted technology for low-literate smartphone users.

Navana has developed a product called Zabaan, which can be integrated into any app with a single line of code and allows businesses to enable human-like voice guidance in any Indian language along with visual cues on top of the app. With Zabaan acting as a voice guide, users with limited digital literacy can browse, explore and transact on sites and applications.

Another solution of Navana allows e-commerce platforms to become more accessible to users with limited or no literacy. So, someone who is unable to read the rate list on an online grocery store can instead interact with a voice-enabled and image-supplemented chatbot that will ask the user, '*Aaj aap kya khareedna chahenge* [What would you like to buy today]?' When the user replies with '*Mujhe bhindi chahiye* [I want lady finger]', the chat bot's complex AI will recognize the words and match it to the product, quickly listing 'bhindi' or 'lady finger' on the menu. The intuitive technology then allows customers to use a combination of touch and voice to build their shopping carts.

SYSPIN and RSPIN

The Indian Institute of Science (IISc), under the projects SYSPIN (SYnthesizing SPeech in INdian languages) and RESPIN (REcognizing SPeech in INdian languages), is collecting data and building models intended to encourage the development of text-to-speech and speech recognition technology in nine Indian languages to drive progress in the fields of agriculture and finance. Through the open-source model, the content is made freely available, which means that anyone can access and use them without incurring any cost. This can be especially beneficial for organizations or individuals with limited financial resources. Additionally, the open-source model encourages collaboration and transparency, which can lead to the creation of higher-quality and more innovative products.

Bhashini

Bhashini is an initiative of the Government of India that aims to use natural language technologies to enable a diverse group of contributors, partners and citizens to overcome language barriers and promote digital inclusion and empowerment in an independent India. Through the creation of language data sets and AI technologies, Bhashini aims to make it easier for people to access the internet, information and apps in their language, including through voice-based methods. It also aims to support the search and viewing or listening in one's language of videos, audio and documents created in other Indian languages, as well as improve access to specialized services in fields such as healthcare, engineering and law. Bhashini aims to enable communication in one's language with speakers of other Indian languages and to provide digital learning in children's mother tongues. To achieve these goals, Bhashini is establishing large open-source data sets and models by bringing together all contributions, both institutional and citizen-generated, in a shared repository that fosters innovation. It also hopes to encourage the development of innovative products and services in Indian languages using this open repository of data sets and models.

Chapter Two

Police 100, Fire 101, Bank Balance *Batao* (BBB) 222?

Roopvati Devi is a farmer who lives in rural Madhya Pradesh. She is a widow who manages the entire farm by herself while single-handedly raising her son. She managed to put her son through school and the small inter-college in the next town. With some money that she had put away, she has also managed to send her son to the big city to pursue a course in IT. It hasn't been easy for her, but Roopvati is determined that her son will live a better life than her. This means that she has to cut corners to ensure her son can survive in the expensive big city.

Roopvati's son Suresh has been living in Satna with a few other boys in a single-room setup. He has been studying assiduously for the last few months, trying hard to complete his course and land a job in a bigger city. He wants to take some of the burdens off his mother's shoulders. Suresh tried to get a part-time job when he started his studies in Satna but couldn't find one because of the COVID-19 situation affecting the market. So instead, he settled for living a frugal life. However, he does need money every once in a while. His coaching classes require frequent payments. There are other incidentals too. He tries not to bother his mother for money, but sometimes he just has to.

In March 2022, India's finance minister announced that every woman with a Jan Dhan account would receive a direct transfer of Rs 500 for three months. Suresh read about this in the newspaper and called his mother to ask if she could transfer some money to him. He owed his roommate Rs 200.

Back in the village, Roopvati had no idea about this direct credit from the government. She only heard about it when Suresh called. She wasn't sure if the money had actually reached her account, so she trudged 3 km one way to check her balance at the bank. On the first day, she didn't even manage to make it on time. By the time she finished her household work and reached the bank, customer hours were over. The next day she tried again and made it to the queue but heard the teller's gruff voice telling another woman two spots ahead of her to check the SMSes she receives from the bank to figure out her balance instead of physically going all the way to waste everyone's time. Unnerved, Roopvati left the queue and walked away.

That night, Suresh called Roopvati to ask if she'd managed to make the transfer, and she told him about her trip to the bank. Suresh assured her that the bank teller was right. All she needed to do was go through the inbox on her phone and get to the bank's SMS. She didn't need to read the text either, which Roopvati, anyway, can't since she has limited literacy. All she needed to do was read the figure, which she can do—she does have some numeric literacy. Roopvati tried, but she was unable to toggle through her phone. With great difficulty, she managed to get to the inbox but couldn't locate the correct text message. She, in fact, inadvertently deleted a few messages.

Not far from where Roopvati, in the village centre, sits a business correspondent (BC) by the name of Rakesh Shekhawat. Rakesh runs a small store that sells mixed goods: some grocery items, vegetables and even cell phone recharge cards. But even that was not enough for his growing family. And so, in 2019 almost a year before therefore the pandemic started, Rakesh decided to become a BC. He'd done this with the intent of augmenting his income. And so, every afternoon, when most shops shut down for a few hours, Rakesh sits under the Big Chaupal[1] and handles queries from people about bank

---

[1] Generally, a central place within a village in India where people gather to socialize and conduct meetings.

accounts, digital transfers, etc. He has been able to make some money. But not quite as much as he had hoped either.

What is a Business Correspondent?

A Business Correspondent (BC) is an intermediary appointed by a bank or financial institution to provide banking and financial services in areas where physical bank branches are limited or non-existent. BCs play a vital role in extending financial services to underserved and remote populations, promoting financial inclusion.

BCs can be individuals or entities like NGOs, post offices, self-help groups, microfinance institutions, and retail agents. They act as representatives of the bank, offering basic banking services on its behalf. These services include opening bank accounts, accepting deposits and withdrawals, facilitating fund transfers, assisting with loan applications, and providing payment services.

To deliver these services efficiently, BCs often leverage technology and digital platforms. They may utilize mobile banking applications, point-of-sale (POS) devices, biometric authentication, or other suitable methods to carry out secure transactions.

The appointment of BCs helps banks reach areas where establishing physical branches may not be feasible due to cost or infrastructure limitations. By using BCs, banks can extend their services to remote villages, underserved urban areas, and other financially excluded regions.

BCs earn income through commissions and fees for the services they provide on behalf of the bank. The compensation structure varies based on the types of services rendered, transaction volumes, and the agreement between the BC and the bank.

Unfortunately for him, around 30 per cent to 40 per cent of the customer queries he handles in a day are only about checking bank balances. And these are transactions that do not pay him a commission. The transactions that give him commission are few and far between. This is why, when Roopvati landed up at the Chaupal two days later, although a little miffed at yet another bank balance inquiry, Rakesh went ahead and checked her balance, nonetheless. Only to find out that the benefit had not actually been transferred yet.

That night, Suresh called again and Roopvati told him that she had been able to check her balance but that she hadn't received the money yet. Suresh then informed her that she didn't need to go all the way back to BC the next day. All she needed to do was call the bank's number and follow the IVR. He coached her on how to do it and what numbers to press for which instructions.

The next morning, Roopvati tried to call the bank. It was a big deal for her—something she'd never done. She finished her work and shut herself quietly in the house before dialling the number. She listened to the recorded IVR with the same reverence with which she would listen to a real person. She got the first one right about choosing the language options but began to get flustered when the options started increasing. Finally, she gave up and disconnected. It's far easier to listen to the BC at the Chaupal!

Not far away, in the neighbouring town, lives Jayaram, who works as a teller in the government bank. Each day, his branch manager reminds him that he needs to open new accounts, which he is unable to do because he spends a lot of time on what he calls 'non-value-added work'. On most days, he gets tired of informing people about their bank balances. He tries directing them to a machine next to the ATM where they can update their passbooks. But they prefer coming to him. He has tried explaining to his manager that if there was another system allowing account holders to check their balances without wasting a teller's bandwidth, he would actually be able to do the job of a banker. He'd be able to open more accounts, sell more loans, track payments and do everything else that he had signed up to do when he became a banker. But his manager reminds him that everyone who walks in is a potential customer.

Meanwhile, Roopvati made her fourth visit to check her bank balance and the second one to Rakesh. A week had passed since the day Suresh asked her for money. That day, Roopvati finally found out that the money had, in fact, been credited to her account. She immediately asked Rakesh to transfer the money to Suresh. In the next five minutes, she promptly received a call from Suresh, confirming that he had received the money too. This technology was beyond Roopvati. She couldn't fathom why it took her a week to find out if she had the money but mere seconds for the same money to reach Suresh, who lives three hours away. She thanked Rakesh profusely. Rakesh, for his part, was grateful for the small commission that he would make and hoped that more clients would come to him for transfers rather than balance checks.

If merely reading about these events from Roopvati's life over a week managed to overwhelm you or make you impatient, imagine what it might be like for Roopvati, for whom this is a lived experience. Roopvati and countless others don't have the wherewithal to do something simple like checking their balance, and in fact, a recent study found that only around 41 per cent found out they received the Jan Dhan transfer via SMS alerts, and around 52 per cent visited a bank, BC, e-Mitras or e-service centres.[2] In your life and ours, this entire episode would've probably lasted thirty seconds.

---

[2] Kejriwal, Saahil, 'Is Jan Dhan Money for COVID-19 Relief Actually Reaching People?', India Development Review, August 5, 2021. https://idronline.org/is-jan-dhan-money-actually-reaching-people/, accessed on July 27, 2023.

Many of us know that we can dial 100 for the police and 101 for fire emergencies. These occur rarely, but when they do, we need to reach out for help quickly and reliably. The poor face a different kind of emergency, and they face it far more frequently. They often need to know how much money they have and whether it is safe. For this reason, they like to keep their cash at home, not in a bank, where it earns no interest and faces the possibility of theft, loss and even a temptation to spend it. They are aware of these risks, yet they choose to keep the money close by. There are other risks they are unaware of: keeping cash at home builds no financial history and results in financial exclusion at many levels—they get no credit and no access to insurance and other financial services.

Let's say you had an emergency and needed to know your bank account balance. You would pull out your mobile phone and look for the last SMS alert from your bank. Or you'd open your mobile banking app and know your account balance in less than half a minute, and that too without any cost to you.

Bank branches like the one Jayaram works in spend countless hours each year informing people of their account balances. Like Jayaram knew intuitively, these hours could be otherwise used for revenue-generating activities and customer redressals. Using a very conservative approach, we at the Digital Identity Research Initiative (DIRI) at the Indian School of Business (ISB) have estimated that the cost of these balance check transactions to the banking sector is around Rs 2500 crore (approximately $300 million) each year.

We propose a simple solution to this very expensive problem: a single toll-free number, 222, across all banks. When dialled from their registered mobile number, this number would only tell the caller their account balance in their preferred language.

The technology and infrastructure required for this already exist in the market. Most banks have a number that, when given a missed call, will send an SMS to the caller with their account balance. However, since each bank does this independently, most people would not remember what that number is. (Do you, for example, know that your bank offers this service? Do you know what that number is?) Moreover, Roopvati is not technologically savvy enough to read and decode an SMS. But everyone can remember BBB, which is 222 on the phone for Bank Balance Batao.

So, the next time Roopvati needs to know her account balance, she can simply dial 222 and listen to her account balance in her preferred language, Hindi, while in the background, the 222 service relies on the UPI 123 infrastructure to fetch the account balance from the concerned bank. When she gets a cash transfer from the government, Banks could also do an automated call informing her about money being credited to her account. This would result in Roopvati and millions of other people like her being able to check their account balance anywhere, anytime, without having to travel to a BC or a bank.

Also, Rajesh will be able to perform more commission-earning transactions and not think of quitting his role as a BC. And perhaps the next time you visit a bank branch with your complaints, the teller may not be busy telling people their account balances while you are waiting in line.

Furthermore, the most important reason why the poor are reluctant to use online banking services is that they fear their money is not safe. A quick and easy way to check their balance, anytime, anywhere, particularly after making a transaction, without any cost or delay, will go a long way towards encouraging the poor to transition to the new digital world that you and I take for granted and benefit from every day. After all, for poor people, even the thought of losing their hard-earned money is perilous.

Information Box 1: Experimental Evidence from India[3]

Researchers affiliated with the National Bureau of Economic Research (NBER) studied an experiment to examine the effects of a transparency mechanism on low-income Indian women who had recently opened bank accounts as part of India's financial inclusion agenda, initiated in 2014. The transparency mechanism in question pertained to the use of voice calls to confirm recent transaction history and bank account balances. The researchers wanted to determine how this intervention impacted the knowledge, trust and account usage of these women, who were receiving government-transferred funds into their accounts. This study was part of a larger effort to understand the barriers to financial inclusion and develop strategies to overcome them.

At the start of the experiment, which lasted for almost a year, women who wanted to check their account balances or make deposits or withdrawals had to spend a significant amount of time travelling to a bank and waiting in line to complete their transactions. This made it difficult for them to access their accounts on a regular basis, resulting in fewer visits to the bank, except when absolutely necessary. To address this issue, the researchers worked with a large public sector bank to design a voice notification service that would help women monitor their account balances and receive a notification when government-initiated payments arrived in their accounts.

---

[3] Field, E.M., Rigol, N., Troyer Moore, C. M., Pande N., Schaner, S.G., 'Banking on Transparency for the Poor: Experimental Evidence from India', Working Paper Series, July 2022, https://www.nber.org/system/files/working_papers/w30289/w30289.pdf, accessed on July 21, 2023.

The service used automated voice calls to deliver updates that were easy to understand, even for those with limited literacy. The voice notification service sent messages recapping bank transactions one to two days after they occurred. In weeks with no banking activity, women received calls summarizing their balance and noting that no transactions had taken place. The service was very popular, with 79 per cent of the treatment group signing up and 70 per cent of the automated voice calls being answered, which is noteworthy given that people often screen calls or are not available to answer them. The voice notification service helped to improve the accessibility and transparency of these women's bank accounts, making it easier for them to manage their finances and stay informed about their account activity.

Overall, the study found that voice notifications improved women's access to account information and increased their trust in bank kiosks. Women in the treatment group were 16 per cent more likely to use phone calls to check their balance information, 7 per cent less likely to rely on the kiosk operator and 6 per cent more likely to know their account balance. At the same time, women in the control group reported significantly higher levels of trust in both the accuracy of the information provided at the kiosk and in keeping their savings at the kiosk. Overall, the use of voice notifications helped to increase transparency and improve the accessibility of these women's bank accounts, making it easier for them to manage their finances and stay informed about their account activity.

These results of the study are encouraging in several ways. A controlled experiment successfully provided reliable information to the women. The women actually used the service, which improved their trust in the banking system. As a result of this trust, more women used other banking services later on, becoming more financially 'included'. The data suggests that the voice notification service was an effective way to increase transparency and improve access to information, leading to more informed and engaged users of the banking system.

Chapter Three

Simple User Interface

Recall our story from an earlier chapter on Voice about Bhishma, who went to the ATM to withdraw an amount of Rs 1000 (approximately $13) but accidentally ended up entering an extra zero, because of which he withdrew nearly his entire savings of Rs 10,000 (approximately $122). Bhishma was then stuck for a few days until he had help from someone to deposit the money back into his account.

Stories like those of Bhishma's are, in fact, not rare at all. There are innumerable cases where the inability to navigate the technology overwhelms the user at the ATM. Withdrawing extra money is just one of the possible mistakes. Sometimes, even semi-literate users or users with limited digital literacy may end up so confounded by the options on the screen that they may drop the entire transaction midway and return later when they have help.

One possible way of avoiding such accidents is by making ATM kiosks entirely voice enabled. While it's true that most ATMs come with a voice configuration, they are usually limited to opening and closing instructions. What if a voice had told Bhishma, 'You have entered Rs 10,000. Are you sure you want to withdraw Rs 10,000?'

This is to say that the chances of a customer responding to a voice-enabled ATM are much higher than when he's transacting on a non-voice-enabled ATM. In the case of an ATM transaction, the aid would be more auditory in nature but would ensure similar results with fewer mistakes.

We have about 2.5 lakh ATMs installed across the country. Naturally, most of these are concentrated in high-density areas or urban areas. Even in urban areas where literacy rates, as well as digital literacy rates, are higher, there are a significantly large number of BOP users who continue to go to branches or to BCs instead of actually using technology-enabled self-help systems, like an ATM.

In fact, this phenomenon is not limited to BOP groups alone. There is a sizeable number of people who are at the top of the financial pyramid in terms of means and wherewithal, are literate in every sense, and can read and write perfectly well, yet they are unable to navigate technology. These are usually the elderly and senior citizens who were perhaps late to the technology train by a generation or two and never got around to adapting to it. If you look around, you will find enough instances within your acquaintances. The elderly, who are unable to go to an ATM and transact entirely on their own, end up going to branches even for simple things like cash withdrawals. To avoid the hassle of multiple visits, they end up withdrawing large sums and keeping them at home.

Sridhar is a sixty-eight-year-old retired schoolteacher who missed the Internet wave by a couple of decades. He was nearly fifty when the Internet Revolution came to India, and by then, he was rather set in his ways. He lived in a middle-class neighbourhood then, raising his two children along with his wife, who was also a schoolteacher. They were comfortable and reasonably well-off. In the 1990s, Shridhar would transact by making weekly visits to the bank branch, where he was given a token—a round metallic disc—that his son loved to play with when he accompanied his father to the bank. After taking the token, Sridhar would sit on a chair, waiting for his turn. When his number was called out, he would go to the counter and either deposit cash, which he had earned from giving extra tuition, or withdraw, depending on his need. Sridhar was set in his ways and refused to change when ATMs started to proliferate.

It's not as if he hadn't tried. He had on more than one occasion. In fact, in the early days, he withdrew cash a few times with the help of the security guard at the ATM. But he never quite liked doing it and never got used to doing it on his own. Over the years, his children grew up and got themselves white-collar jobs. Sridhar's family moved into their apartment in a more upscale part of town. His children, of course, are far savvier than Sridhar or his wife are. They tried on many occasions to get Sridhar to become 'more independent', as they called it, by trying to use ATMs himself and even going digital on his phone.

But Sridhar has actively resisted graduating to technology. He prefers his old ways of going to the branch, withdrawing money, smiling at the staff who have aged with him, and transacting with humans instead of machines. He distinctly remembers the few occasions when he had tried to operate the ATM on his own but just couldn't seem to get it right. For the teacher in him, it was a particularly uncomfortable situation to be in. So, he gave up technology entirely in favour of a method that actually worked for him.

Despite his aversion to technology, Sridhar does operate a smartphone without much difficulty as he uses the 'Easy Mode'. When one chooses this mode, the user interface changes completely. The numbers and the fonts become larger and easier to view. Most unnecessary apps and settings disappear entirely from the screen, making it look simpler and more manageable. This makes the elderly who are using the phone less stressed about navigating several windows and several apps. In some smartphones, this mode also automatically activates an audio interface as well, where, in addition to seeing things on the screen, the user can also hear a voice that reads the screen and guides them on what to do. Therefore, the common ailments that come with ageing, like poor sight and hearing, also get automatically addressed.

ATMs already have plug-in options for visually impaired people, and many even come with a Braille panel on the side. What if, in addition to these facilities, ATMs also had smartphone-like easy mode functions? Imagine if Bhishma or Sridhar could activate the mode and only see a few options that they needed along with some helpful images rather than having to choose their preferences from six or even eight options on the screen. Would that not be a simpler way for them to transact, and would that not enable banks to draw more people into self-help practices?

Ujjivan Small Finance Bank has worked in this very area to improve the interface between customers and banks. Their example is a useful case study on simplifying digital transactions for groups with limited digital literacy by simply working on the user interface.

How Ujjivan Made Banking Apps Simple

In recent times, Ujjivan Small Finance Bank, which works with the goal of including this very group in financial systems, has started to use voice as a critical tool in its Ujjivan Mobile Banking application. Ujjivan offers a range of banking services through a voice-guided interface in eight regional languages. This app is particularly noteworthy because it is the first of its kind in India to use artificial intelligence and machine learning to enable voice search and assistance, making it capable of understanding not just languages but even different dialects within a language.

The use of voice technology in the Ujjivan Mobile Banking app has several benefits for users like Bhishma and Sridhar, who may not be comfortable with traditional, text-based interfaces.

For example, the recorded audio guide on every screen provides an additional layer of guidance and support for account holders who may have limited literacy. The app's visual representations of physical banking transactions through images, such as paying an EMI or depositing money at a bank, also help to bridge the gap between the physical and digital worlds and make the app more accessible to the semi-literate. So, Bhishma, whose reading speed may be very limited, can look at an image and immediately know what that option could be.

Most users who have limited digital literacy also end up only partially using all the features and benefits that their bank account comes with. However, because Ujjivan's technology enables end-to-end voice guidance and chat-bot assistance with voice search, these users are able to access more services offered by the bank.

By offering a voice-guided interface capable of understanding different dialects, the app can reach a wide range of BOP users and positively impact their financial lives. In some ways, the Ujjivan Mobile Banking app serves as a model for the potential of voice-based technology to improve the access and inclusion of financial services for all individuals, regardless of their technical skills or familiarity with traditional banking methods.

Chapter Four

Microangels

The practice of lending can be traced back to ancient Babylonia. As early as 3000 BC, farmers would 'borrow' seeds from a lender and cultivate them to pay back the principal. Far away in ancient Rome and Greece too, money lending was fast becoming an institution to be reckoned with. The practice differed between continents, as did the rates of interest and the penalties for default. Though separated by several seas and many centuries, the practice continued to flourish in these civilizations much before banks or any form of banking could be legalized.

These moneylenders were essentially individuals operating out of their own free will and industry. There was no formal network connecting them. Despite this, they've stood the test of time because they fulfil a very basic need—that of providing funds when all other methods have failed for the borrower. Even though formal banking structures emerged, and people could borrow from banks at lower rates in addition to saving with them, these independent moneylenders continued to be important. And continue to be so, even today. There are several reasons for this:

Loans provided by banks can be quite cumbersome to apply for. They require a lot of documentation and time, and rejections are frequent. Often, large loans require collateral. Sometimes, banks are just not interested in providing modest loans because borrowers are too small and too poor to serve profitably.

None of this is necessary with the moneylender. There is no paperwork, and the loans are usually immediate in nature—just the need itself. The flip side, of course, is the high rate of interest, which is how moneylenders hedge their risks. In fact, much of the argument against individual moneylenders revolves around the unethical means they sometimes deploy to ensure borrower compliance. Also, most moneylenders cultivate their network of borrowers over the years. Each of them is known personally to the lender. A new borrower is most often introduced and vouched for by an existing one. Even today, moneylenders continue to be the only resort for the poor who are not a part of the financial machinery when they need money.

Stories of unscrupulous moneylenders can be found in perhaps every kind of literature and folklore. They were often portrayed as wily men who were lying in wait for poor peasants to fall into bad times. Shakespeare's *Merchant of Venice* immortalized Shylock as the greedy moneylender out to get Antonio by any means, and when he does manage to get the merchant into a bind, he literally demands a pound of his flesh; such was his anger towards the gentle Antonio, who lent money interest-free and ruined the trade for everyone else in Venice.

Closer home in India, Munshi Premchand's books, which were renowned for their realism, almost always painted a picture of rural India and the ravages of poverty. None of them was complete without the village moneylender and his unwavering dedication to retrieving the loan from his debtors, come what may.

One of his most profound books, *Godan*, talks about the family of Hori Mahato, which is reduced to complete penury because they accumulated insurmountable debts trying to buy a cow. Even though they eventually 'sell' one of their daughters into marriage, the money is insufficient to bail Hori out, and he dies before he can fulfil his only desire of buying a cow.

This is not to say that moneylenders were intentionally villainized. The truth is that moneylenders were, in fact, a class that profited from the poor's inability to become entirely debt-free. Keeping them bonded to loans that continued to yield high rates of interest was important to lenders, and they freely exploited the relationship to that end.

The moneylenders operate differently now, as compared to the times in which Premchand wrote. A hundred years ago, lending was the primary occupation of the moneylender. In contemporary India, moneylenders are often people keeping the tradition alive within the family. Some of them may even have jobs, with money lending being something they do in addition to their primary occupation.

How Do Moneylenders Operate?

Subu is fifty-five. He lives in a small town in Tamil Nadu with his wife and two teenage daughters. He works in the local fertilizer factory, which is the largest employer in his small town. He lives in company-provided quarters and rides a motorbike—a huge improvement from his childhood when his father rode a bicycle. By all means, Subu and his family have a comfortable life.

By morning, Subu works as a supervisor in the factory, but come evening, he heads out for his daily round-up of debtors. Subu is a moneylender in that small southern town where banks exist, but so do enough white-collar people who prefer borrowing from individuals like Subu. In fact, Subu's clientele includes many contractual labourers employed in the very factory where he works.

Money lending was not Subu's original plan. His father had been a lender, and that's all he did. In fact, Subu got an education when most of his contemporaries didn't because his father could afford it. But when Subu's father passed away suddenly in a train accident, he had to put his father's shop in order. It took him months to go through the meticulously maintained notebooks and track each of the debtors on the list.

By the time he was done, he realized he was in the thick of it. People were already borrowing from him. A relationship his father had started and so assiduously cultivated could not be put to an end just like that.

And so, it came about that Subu just accepted his fate and became a part-time moneylender. Now his clientele expects him to drop in once a week, even if they're not due to pay. Subu sits for a while, has a cup of tea and inquires about the family before going on his way. It is in these meetings that he is sometimes introduced to potential borrowers as well. Word of mouth is the best testimonial in a business such as this, and Subu has gotten used to rich rewards by now. He likes the idea of a comfortable retirement. Since he doesn't have a son, he must plan for his wife and himself.

Not all moneylenders are like Subu, though. Rama Rao, sixty, operated in the same vicinity as Subu, often with overlapping clients. Since rates of interest are at the whims of the lender, Subu and Rama Rao competed to gain market share. Rama Rao would offer marginally lower rates in an attempt to nudge Subu out but would use all kinds of force to ensure repayment. No delays of any sort were tolerated.

Eventually, Rama Rao fell ill and started winding down his business. There is an opening in the market now, and any day a competitor might emerge and challenge Subu's temporary and peaceful monopoly.

Why Do Moneylenders Exist?

Despite players like Rama Rao, it is difficult to outlaw, ban or over-regulate an activity that has such a robust demand and serves both important and legitimate needs.

Take Raji's story, for instance. Raji is a contractual labourer in the fertilizer factory, where Subu is a supervisor. She is well aware of the fact that he's a moneylender. Some of her neighbours and co-workers have borrowed from him. A year ago, she tried to take out a bank loan to help finance the wedding of her older daughter. With great trepidation, she walked into the bank's branch and waited till she could muster the courage to ask the first available staff member about her loan. She navigated the intimidating system of generating a token, keeping a lookout for her turn and finally reaching the counter to place her request. She found that the bank employees were not particularly polite or welcoming, which she was prepared for. What she wasn't prepared for was the barrage of questions: why did she need a loan? How much did she need? Would she be able to pay it back? Did she have collateral? What documents was she carrying?

After waiting for over half an hour, Raji was told to come again with her photos, Aadhaar card and documents showing her income. Being a contractual labourer, Raji did not have formal pay slips. So, she began making rounds of the factory to acquire some form of documentation. It took her a week or so to get it. When she returned to the bank, the officer informed her the documents from the factory were unacceptable as she was paid in cash, and they had no way of verifying her income details. The officer then offered the option of taking a loan by using her gold jewellery or land documents as collateral. Her husband had passed away some years ago, and no official paperwork had been done for the small land parcel he'd owned. As for the gold jewellery, she owned only one gold chain, which she had planned on giving to her daughter as a wedding gift but thought she could try and convince someone in her family to lend her their jewellery to use as collateral.

Meanwhile, this back-and-forth was costing Raji a lot of time. The wedding date was approaching, and the groom's family was pressuring her to start some of the rituals. Finally, after a particularly challenging day trying to find someone in her family willing to give her some gold jewellery to use as collateral, she finally gave up.

She walked straight to Subu's house only to find that he and his family had gone to the city for a week to attend a wedding in the family. It was, after all, the wedding season. Left with little choice, Raji went straight to Rama Rao's house, where he was quite pleased to see her. She was offered water and tea and given a place to sit in the shade. Rama Rao already knew about Raji's daughter's wedding plans. He offered her a loan on the spot. Fifteen minutes and a few thumb prints later, Raji was holding the much-needed money in her hand, the same money for which she'd made so many unsuccessful trips to the bank. Raji was well aware of the frightening interest rates that she faced, but she needed the money *now*. She would worry about the interest after the wedding was over. She told herself that she had tried to take a loan from the bank at better terms, and it was not like she wanted to pay more interest.

Raji's relief soon gave way to excitement. She spent the money well and gave her daughter as lavish a wedding as she could. Content that she'd done well for a widowed mother, a month later, Raji had started to pay off the exorbitant instalments of her loan from Rama Rao.

A month later, she bumped into Subu at the neighbour's house. Subu, like Rama Rao, knew everything about the loan situation already. He gave her good wishes but also scolded her in the most paternal manner for not reaching out to him on time. He offered to pay off Rama Rao's loan and have it transferred to himself at a lower rate.

For Raji, this was a huge and unexpected gift. She agreed immediately. She probably knew that even Subu's rates were higher compared to the banks, but his recovery methods were nowhere near as exploitative as Rama Rao's. And so Raji got her loan transferred from Rama Rao to Subu.

What about Microfinance Institutions?

One might wonder that a more useful response by society to the moneylenders would be to cultivate less onerous and more ethical means of serving the financial needs of people who need loans. The microlending movement and the rise of microfinance institutions (MFIs) in the last three decades have tried to respond to this.

Muhammad Yunus and Grameen Bank received the 2006 Nobel Peace Prize for pioneering the Microfinance Revolution. Visionaries such as the late Sir Fazle Hasan Abed of BRAC (Building Resources Across Communities) and John Hatch, the founder of FINCA and the creator of Village Banking (not surprisingly, Grameen also means Village) were also the forerunners in this revolution. Though the word microfinance does encompass micro-savings and micro-insurance, the focus of the initial efforts for nearly two decades in the 1980s and 1990s was on microlending that involved microloans as small as $50 or $100 (Rs 4000 to Rs 8000).

The Yunus model focuses on providing loans to low-income individuals and businesses that may not have access to traditional forms of credit. These loans are typically small in size and are used for a variety of purposes, such as starting or expanding a small business, financing education or healthcare, or meeting basic needs.

Yunus' model is based on the belief that even the poorest individuals can improve their lives and lift themselves out of poverty if they have access to the right resources and support. It emphasizes the importance of providing loans with low-interest rates and flexible repayment terms, as well as providing education and support to help borrowers succeed. The model has been successful in helping millions of individuals and businesses around the world access the financial resources they need to improve their lives and build better futures.

What is remarkable about the various microlending models is the fact that the small loans are made to entrepreneurs without any physical collateral, and despite the lack of physical collateral, the repayment rates have been unusually high—over 95 per cent in many cases.

Despite such high repayment rates, it has proved extremely difficult to create self-sustaining and economically viable models of micro-lending that do not rely on some form of explicit or implicit subsidy by governments, charitable organizations and socially conscious individuals willing to sacrifice their time and talent for lower levels of compensation as compared to potential earnings in other jobs.

The question is why, despite such enviably low default rates, the microfinance evolution's success has remained tepid, notwithstanding a meteoric rise of microfinance institutions (MFIs) since the 1990s, despite which, according to the All-India Debt and Investment Survey 2019, in rural India, less than 5 per cent borrow from MFIs and over 20 per cent borrow from moneylenders.[1] The answer, it turns out, is that microlending is highly cost-inefficient.

Microangels

Co-author of this book, Bhagwan Chowdhry, and his former colleague Amit Bubna from ISB illustrate why the moneylender is such a formidable competitor to the formal financial institutions in a research paper that appeared in the *Review of Finance*, the official journal of the European Finance Association. They argue that a feasible way to ameliorate the

---

[1] Agrawal, N., Ponnathpur R., Seetharaman, S., and Misha S., 'Insights from the All-India Debt and Investment Survey 2019', Dvara Research, February 1, 2022, https://www.dvara.com/research/wp-content/uploads/2022/02/AIDIS-Slide-Deck.pdf, accessed on July 16, 2023.

impact of high-interest rates charged by moneylenders is to generate competition by creating and funding local people who would compete with the moneylender. They develop a model where a bank uses informed local people as intermediaries for on-lending. They argue that in any given neighbourhood, there exist several individuals who possess the information that is relevant for screening and monitoring potential borrowers. Besides information, some of these individuals may also possess (possibly illiquid) collateral, such as houses, jewellery and animals, against which they could obtain loans from various formal financial intermediaries, such as banks, for on-lending to borrowers who may not have similar collateral but are in need of funds.

The model suggested by Bubna and Chowdhry taps into these individuals and recommends turning them into an army of 'Microangels' who can compete with not just moneylenders but also the image of moneylenders as an avaricious and self-serving group.

Ajit Pal, from the village of Azizpur (about whom we shall talk more in the later chapters), would make a very good candidate for a Microangel. With his shop in the centre of the neighbourhood, his flourishing clientele, and his natural ability to listen to people's problems, Ajit would not only know the people really well but also their capacity for borrowing. He probably already sells groceries on credit. The Indian system of keeping hisaab when customers are low on cash is age-old and, in fact, encouraged by traders and sellers to ensure that the business keeps running. It's a tool for retention as well. Saying no to a regular customer just because they don't have the money immediately is not a sound business practice. This is also where the goodwill of both parties comes into play. Ajit has a long-standing relationship with his customers. He knows they will pay him back like they always have.

A Microangel like Ajit is in a unique position. He is often roughly aware of his customer's income, assets and liabilities, either because they tell him directly or because he hears about them from his network of family and friends. He also can tell from his customers' buying patterns when they are in need of funds, and he can upsell his loans to them (and possibly even save them the embarrassment of asking). In good times, he will anyway come to be known as much for his moneylending services as an intermediary as he is now for his grocery shop. It can consolidate his standing and his image. He is not just an unscrupulous moneylender but a regular grocer who also lends.

But these potential lenders, the Microangels, may lack the necessary enforcement 'technology'[2] that is available to the moneylender. Even when Ajit can credibly threaten to deny further credit to a borrower who defaults on a loan made by him, it is ineffective in deterring voluntary default in the face of competition among such Microangels. Following the default against Ajit, the borrower may approach another Microangel for a loan subsequently. The possibility of voluntary default by a borrower in the presence of multiple Microangels without an effective enforcement strategy reduces (and may eliminate) each Microangel's incentive to offer a loan in the first place. The availability of multiple credit sources provides borrowers with an incentive to default voluntarily, making the bank's on-lending mechanism a non-starter.

Bubna and Chowdhry consider a coalition in which each Microangel commits to refuse credit to a borrower who may have defaulted against another member of the coalition. If such an arrangement can attract Microangels as lenders, it would reduce the borrower's incentive for voluntary default. The

---

[2] They are often unethical and sometimes even violent.

incentive for a Microangel to join the coalition would be the possibility of making a loan profitably; in the absence of such a coalition, that is not a possibility. However, each Microangel would continue to compete against other coalition members for the borrower's business before any default.

This coalition strategy resolves the Microangels' commitment and contract enforcement problem. It is worth noting that the primary purpose of the coalition here is not necessarily as an institution, such as a credit bureau, for sharing borrower default information. In fact, even if such information were publicly available, lending, in the model, would not take place in the absence of a coalition. The Coalition is a strategy by which lenders credibly commit not to lend to a defaulter, even when it may be individually desirable to do so after the fact. Would this strategy, effectively limiting the borrower's opportunity to default multiple times, be sufficient to facilitate on-lending? Not yet.

They argue that when the moneylender is the cheapest producer of loans, the Microangels, even if they form a coalition, cannot outcompete the moneylender who possesses superior enforcement technology. The moneylender can repeatedly provide loans to the borrower, effectively shutting the Microangels out of the market for microcredit. The monopoly moneylender with superior enforcement technology can outcompete the Microangel coalition because the moneylender also enjoys the lowest transaction costs of lending. They show that a credible competitive threat to the monopoly moneylender can only arise if the Microangel coalition can also be made cost-effective, either by direct subsidies or by measures such as standardization, economies of scale and the implementation of best practices. They argue that franchising is one potential mechanism that could deliver both cost efficiencies and the ability for Microangels to form a coalition.

How Can We Franchise Microangels?

The recent advances in technology and the widespread adoption of a digital identity such as Aadhaar[3] would make it possible for banks and financial institutions to franchise through Microangels in a cost-efficient and profitable manner while also reducing the cost of borrowing for the people at the BOP. In the previous section, we talked about the Microangel model proposed by Chowdhry and Bubna, and now we will attempt to illustrate through a series of examples how this could be potentially implemented in India.

Let's say the CEO of ABC Bank wants to increase the offering of unsecured microloans to the BOP customer segment in Azizpur, the village of Ajit Pal. The bank could decide to go through the traditional channels of increasing their loan portfolio by lending through their branches or through priority sector lending to microfinance institutions for on-lending, where the interest rates for the customer could be anywhere from 20–40 per cent. Alternatively, the bank could ask Ajit and Mohan, the village sarpanch, to become Microangels and act as a franchise for the bank while providing the enforcement technology through a mobile app and also offering to loan money to Ajit and Mohan at an attractive rate of 11 per cent in exchange for some collateral from them that is to be used for on-lending to customers in the village.

Ajit decides to sign up and become a Microangel and being a kirana (local grocery store) owner, he has very little excess liquid cash available with him that he can use for lending to the villagers, so he decides to use his shop ownership documents as collateral and borrow Rs 3,00,000 (approximately $3658) to

---

[3] Chowdhry, Bhagwan, Amit Goyal, and Syed Anas Ahmed, 'Digital identity in India', *The Palgrave Handbook of Technological Finance* (2021): 837–853.

begin lending from his shop. Mohan too, has decided to sign up and become a Microangel, but since he has enough savings with him, he decides to only use the technology offered by the bank and, in exchange, pay 1 per cent of all loans disbursed by him as fees to the bank.

The bank provides them with technology in the form of a mobile application where they are able to register customers and enter details such as loan amount disbursed, duration of the loan and repayment status. So, for example, when a customer decides to default on a payment with Ajit and get a new loan from Mohan, the mobile application will inform Mohan about the previous default and advise against giving a loan to the customer. The bank also allows Ajit and Mohan to decide on what interest rates to charge different customers based on the local information they have about those customers.

Franchising is a mechanism that would allow banks to simultaneously use informed Microangels and address the market failure resulting from the risk of strategic default. It is equally important that franchising ensures that the Microangel with the lowest transaction cost for a particular borrower is sufficiently cost-effective to compete with the moneylender and attract her business, resulting in the natural segmentation of borrowers based on the smallest lending costs.

In the Bubna–Chowdhry model, formal financial institutions found it profitable to lend funds to these individuals for two reasons. One, individuals with local information could provide collateral. Two, because they made multiple loans, the size of the loans made by the formal financial institutions to each Microangel could be substantially larger, resulting in reduced transaction costs.

What would happen with the possibility of competition among formal financial institutions in offering franchises? In the extreme case of unfettered competition between banks to offer franchises, strategic default again becomes feasible, and the market for credit through franchising fails, leaving only the usurious moneylender. This suggests that there may be little reason to encourage competition between banks in offering franchises. In fact, some monopoly power—exclusive territorial rights, for instance—for each bank may be a desirable policy.

Chapter Five

Digital Didi[1]

'*Chalo, Didi se poochte hain* [Come on, let's ask Didi]!'

Radhika wakes up promptly at six every morning. Like all the other homemakers in her lower-middle-class neighbourhood, she first fills the water in the overhead tank. She cooks for her husband and two little children, all of whom leave around the same time. Then she quickly gets ready herself before the rest of the household wakes up. She packs their lunches, gives them their breakfast and sees her husband off as he rushes to catch the local train to his job in the heart of the city. Finally, she has her breakfast before heading out to the bus stop with her children.

At the bus stop, she waves goodbye to them and sits down. She slowly takes out a folded cotton waistcoat in a shade of deep vermillion and wears it over her saree with a contented look on her face. She looks up and sees another young girl, in the exact same waistcoat as hers, across the street. The strangers exchange knowing smiles with each other just before someone taps Radhika on the shoulder. She turns to see an elderly woman looking at her helplessly. Moments later, Radhika and the older woman are huddled over the latter's phone. Radhika is speaking patiently, and the old lady nods intermittently. In the meantime, another bus briefly stops in front of them. Unbeknownst to them, a similar interaction is transpiring in one of the seats on the bus. A young college girl, also in the same vermillion waistcoat, is saying something to a middle-aged man as they both look at his phone.

Who is Radhika? And who are these young women in the dress code, you wonder? We'll explain that just a few paragraphs ahead. First, let's take a look at what does literacy mean in the financial inclusion context and two programmes that have tried to solve for this.

The Oxford English Dictionary defines the word 'literacy' as a noun that means 'the ability to read and write'. That is a fairly straightforward description, one would think. But it is not that simple, for example let's take the case of Meetu Roy who can read and write in her mother tongue Bengali and hence is even counted as a literate person in the government census but when she visits the bank and is given a form to read and fill, she can barely understand a few words and requires assistance to fill the form. Here, her ability to read and write was not sufficient to understand the requirements of the form and what she needed was some help in learning about basic banking terms and processes.

Similarly digital literacy has its own shades as we have seen during many of field visits in rural areas where we have come across many women proudly telling us that they now owned a smartphone and use it to communicate with family and friends, browse YouTube and watch movies and listen to news on their smartphones. One may argue that these women are digitally literate yet most of them do not using any digital banking apps or services on their smartphone. When we ask why they do not use any digital banking apps or services they say that they don't understand how these apps work or what is needed to register and use them, and if they can trust these apps.

Digi-Prayas

In 2017, Digital Empowerment Foundation and Axis Bank jointly launched a comprehensive programme called 'Digi-Prayas' to significantly increase financial literacy among individuals living in rural areas of India. The programme targeted a total of 80,000 people living in twenty-four villages across the country and aimed to provide them with a wide range of knowledge and skills related to digital banking.

As part of the initiative, a fully functional digital banking ecosystem was established in the selected villages. This included the deployment of BCs, EDC/POS (Electronic Data Capture/ Point of Sale) devices, and micro-ATMs, which were designed to make it easier for people in the villages to access financial services and carry out digital transactions.

In addition to setting up the necessary infrastructure, the Digi-Prayas programme also included a series of awareness sessions and training workshops. These were intended to educate people about various digital banking options, such as mobile banking and UPI, as well as about government policies related to digital banking. The programme also provided training on electronic payment systems like IMPS and UPI and enabled merchants in the villages to use electronic payment systems like UPI and POS terminals.

The DigiSupport programme was developed as the second phase of the Digi-Prayas initiative, with the goal of increasing awareness about digital financial tools in rural areas of fourteen Indian states. As part of the programme, the target population was encouraged to use digital banking services for cashless transactions. To facilitate this, the programme conducted a range of awareness activities, including programmes on digital financial literacy and internet banking. These programmes were led by local leaders known as DigiPreraks, who use a variety of technology resources, such as tablets and videos, to provide financial education to the community.

Internet Saathi by Google and TATA Trust

The Tata Trusts' Internet Saathi programme, launched in 2015 in partnership with Google, is yet another social endeavour that works to empower rural women in India to serve as agents of change by promoting digital literacy within their communities.

Through this programme, more than 81,000 'Saathis' have trained over thirty million women in more than 2,90,000 villages.

Under the Internet Saathi programme, after providing hands-on training, the Saathis are given two smartphones or tablets and sent to their villages for twenty days a month over six months. They raise awareness about the internet's benefits, teaching others how to access crucial information like weather updates and relief measures. Once a Saathi completes training in her own village, she moves to nearby villages to train more women to use the internet.

The programme did have significant impact on the beneficiaries and the Saathis, for example many of the women who first learnt to use a smartphone have gone on to start business and provide employment not just for themselves but other in their community.[2]

Why Did the Programmes Not Succeed?

Despite the effectiveness of each of these programmes in the pockets they worked in, why are we still saying that they did not succeed? An important distinction we are making is that the programmes did not fail. They just did not succeed in making a huge impact on the digital literacy landscape—the kind of dent needed in order to invigorate a landscape as vast and complex as ours.

Where these well-meaning and, to reiterate, effective programmes fell short was with respect to scale. None of these

---

[2] Chakravarti, Ankita, 'Google Internet Saathi Programme Turns Messiah for Unemployed Women in Rural Areas', India Today, March 8, 2021, https://www.indiatoday.in/technology/features/story/google-internet-saathi-programme-turns-messiah-for-unemployed-women-in-rural-areas-1776789-2021-03-08, accessed on July 16, 2023.

programmes could effectively scale up beyond the thousands or, at best, a couple of hundred thousand.

The point we've been making throughout the book is about solutions being SHUb. While the programme's presence may have been ubiquitous within small geographic areas, it was not widespread enough for the whole country to benefit from it. Could there be one single solution, one single option that has the potential to be simultaneously as ubiquitous in Bhuj as it is in Benares?

We propose exactly one such solution, but before we explain it in more detail, it's important to understand the bedrock of the said solution: the fact that we learn best in one-on-one settings.

We Learn Best in One-on-One Setups

So many of us use self-help kiosks of different kinds on a regular basis without much thought. We buy metro cards and tickets from machines, punch out token numbers on the rare occasion that we visit bank branches, and use vending machines to buy water or snacks in busy spaces. We might sometimes do these with a slight hesitation, but mostly, we manage just fine with a nudge from a standby attendant or someone in the queue around us who has done it earlier.

Think back to the time when you started using your first phone. It was most likely a feature phone with large buttons. From this, you would have graduated to maybe a slightly fancier flip phone with flatter buttons on the keypad. Most of us would have graduated from this kind of phone to a BlackBerry that was more complex and came with additional features. Eventually, BlackBerrys got phased out as well, and we jumped on the smartphone bandwagon.

Now we'd like you to think about this journey that you made and focus on what you learned. Not just how you learned how to use your first phone, which was a big milestone, but how you adapted to each new technology that kept entering the market every couple of years and then eventually every couple of months. If there is one thing technology does, it is change. And the rest of us who are users of technology immediately adapt. This act of adapting is essentially the act of learning new technology. How many of us attended classes on how to use these phones? I'm pretty sure your answer would be that none of us did. We all learned either on our own, by making mistakes, or with the help of someone who taught us. Sometimes, both. Maybe an older sibling, a friend, or, for some of us, even our children.

None of us enrolled in classes or workshops, yet we managed to figure it out on our own. And this is the principle that we use when we suggest the Digital Didi model: We learn best one-on-one. This is essentially what the Digital Didi does. She does not put people in a classroom or a group setting. She teaches you one-on-one.

An important factor to consider in adult learning of any kind is that it puts the learner in a vulnerable position. By choosing to learn, we are also demonstrating our inability to do something on our own. This, for a lot of adults, might not be a very comforting situation to be in. But by learning in the safety of a one-on-one space, this transition becomes somewhat easier.

Digital Didis

Prime Minister Narendra Modi promised a new digital India that will financially include everyone, from workers in the informal sector to women working at home to our ageing parents receiving a pension. The JAM trinity, consisting of Jan-Dhan-Yojana, i.e., bank accounts for everyone, a unique electronic Aadhaar ID, and mobile banking facilitated by UPI are all in place. Indeed, we have come a long way with 80 per cent of the population having bank accounts, 1.25 billion

Aadhaar IDs, and the ever-increasing use of digital payments and mobile banking accelerated by UPI, which has made cash transfers simple, direct and without leakage that was rampant in the past.

Yet, we are far from declaring victory. Just ask yourself how many of you are paying your household help digitally or when we look around, we find that many of our older relatives do not feel comfortable using their smartphones for banking transactions or even digital payments for autorickshaw fares, day-to-day shopping and chores. Cash is still king.

There appear to be two major missing features in financial inclusion: convenience and safety. Even as digital payment apps become easier and simpler to use, they are still not as intuitive as using cash. Furthermore, after a digital transaction, many users are afraid of whether the transaction really took place or not, especially when it comes to receiving a payment in their bank account. Fear of fraud and tampering with one's hard-earned money is a severe deterrent for many.

When we attend conferences on financial inclusion, most participants are sophisticated and understand these issues. The solution that is offered is that we need to create digital literacy. But that is easier said than done. How do we do it effectively enough so that the fears and misgivings many people have, are not merely alleviated but are, in fact, eliminated? We certainly cannot offer classes or group camps in digital literacy.

Back to Radhika, our protagonist in this chapter. Radhika is part of a tribe of Digital Didis. They are not yet a reality, but part of a solution we propose is employing an 'army of young women' and calling them Digital Didis.

Who is a Digital Didi, to begin with?

She is a literate young woman who could be in college or even high school. She could be a homemaker, a freelancer, or someone who runs her own business out of home in beauty, personal care or even food. She could be doing anything as long as she has the required digital literacy and the time to dedicate even an hour or two a day to being a Didi.

What Does She Do?

She's a digital helper that anyone could feel comfortable approaching in malls, on buses and trains, on the street and in stores to help them with any digital transaction. A helper Didi. A trustworthy Didi. A Didi would patiently show and help people how to carry out a transaction and assure them of its safety. She's human and inspires trust.

Who Approaches a Digital Didi?

Absolutely anyone who needs help completing a digital transaction can approach a Didi. A Didi's clientele is gender, age and class agnostic. From literate elderly people who are simply wary of technology to men from the BOP with limited literacy to women who have neither literacy nor digital know-how, a Didi is there for all. There may be people who are fearful of transacting on their own because they worry about losing money. Having a human being present during the translation and, in fact, helping them close it, gives it legitimacy and a sense of closure—that the transaction did go through because a 'person' confirmed it and not a machine. The fact that she is a woman and the fact that she's called a 'Didi' and not a 'Madam ji' or something else formal are both intentional. Her gender evokes trust almost immediately, and calling her Didi, or sister, gives her respect even before she starts to transact for her customers.

How Would the Digital Didi System Work?

Each transaction that a Didi facilitates could be tagged with Digital Didi's identity for accountability and incentives. For instance, each Didi has her own QR code. Anyone approaching her needs to simply scan the code and access the Didi app, where the Digital Didi transacts on their behalf. Didis would be paid a small stipend by banks or financial institutions, and this would be something Didis would do not necessarily full time but in their downtime, as they walk about doing their other daily chores, including going to or back from school or college, shopping or other leisurely activities. She could be anybody and a Didi. She could be a homemaker, a tailor, a factory supervisor or even a teacher opting to be a Didi in her downtime.

The number of hours or even minutes that a Didi works is entirely up to her. Her 'shift' starts the moment she dons her uniform over her clothes and ends the moment she takes it off. It's entirely up to her how long she wants to be a Didi. Some days, she may have a few hours; on others, just ten or twenty minutes.

The use of Didis would reduce the need for opening up many bank branches and hiring full-time employees, and the funds released from these savings could help employ many Didis whose presence will become ubiquitous. Over time, people will begin to know and trust a few Didis who will become their regular go-to Didis for help with mobile financial transactions.

Radhika, for instance, spends an hour every morning helping people before she goes back home to resume her chores. Later in the day, before she has to pick her children up again, she dons her waistcoat again and turns into a Didi. She does it at will, working more hours on one day, fewer on another. When her children are home on holiday, she doesn't go at all. Being a Didi means making some money on the side and achieving a modicum of financial independence. She doesn't struggle to ask her husband for money when she needs it; he is a considerate and kind man, but just having her own money and not having to ask him makes Radhika happy.

Being a Didi keeps her socially active too. She recalls her pre-Didi days, when she only had a few neighbours to talk to. But now, she meets new people every day. She's made friends with some of her regulars, who know her routine and seek her out often. From carrying out an entire transaction for her older clients to showing them how to do it to finally reaching a stage where they just make the transaction and check with her if they've done it right, Radhika has come a long way, and so have some of the women she's helped. In fact, inspired by her, three young girls in her neighbourhood have also turned into Didis. They practise after college hours and on the local trains during their daily commutes.

She likes being called 'Radhika Didi' and the respect she gets. Helping people is cathartic, listening to them and their struggles has given her a newfound appreciation for her own life.

What Happens in the Long Run?

The most effective method by which we learn is imitation. When mobile phones were introduced, no one offered classes on how to use them. People learn how to use them by watching other people. Now mobile phones are being used effectively not only by the rich and middle-class people but even by many poor in villages all over the world. As people begin to build trust by interacting with Didis, many will slowly learn how to carry out digital transactions on their own. Young women who become Didis will not only earn some money on the side, but their experience of successfully helping more and more people will build their confidence and make them stand out for more attractive formal jobs.

The Digital Didi programme makes a very strong case for women's empowerment as well. It gives them financial independence and a sense of worth. Most importantly, it could pave the way for more significant work that Didis go on to do within or even outside the financial system.

Chapter Six

Redressal

Mistakes and Errors Happen. Is My Money Gone?

Ajit Pal has lived all forty-five years of his life in Azizpur village, about 80 km from Bhopal. In these four-and-a-half decades, he has seen his tiny hamlet grow into a village that has a dispensary, a post office and even an ATM within a few kilometres of the main road. Ajit belongs to a farming family but has chosen to run a shop instead of tilling the land. Always known as the ambitious one in his family, no one was surprised when Ajit declared that he did not want to farm. He had always shown signs of industry and wanted to achieve more instead of settling for what he had.

When he started, Ajit converted a lane-facing part of their home into a small shop for knick-knacks. Over the years, he has become a full-fledged grocer. His income grew slowly, and finally, there was some money to put away for a rainy day. Like the rest of his family, Ajit too saved his money in a post office account.

After the launch of the Pradhan Mantri Jan Dhan Yojana (PMJDY) in 2014, Ajit opened an account with a bank. In fact, he has been gradually saving more money in the bank as compared to the post office because he was told that the interest rates are better.

A few years later in 2017, Ajit felt confident about expanding his small business and decided to extend the shop by turning the adjoining room into a storeroom of sorts. He needed money, of course. The option of a bank loan seemed to get more challenging as the days passed. To begin with, he could not fully understand the application process. Then, some of his friends had warned him that 'people like us' are often rejected by banks due to the absence of '*kaagaz*' (documentation) and income proof. So, it transpired that, just like in the old days, he approached the village moneylender for the loan instead.

The investment turned out to be a wise one: the shop is flourishing, and Ajit has been able to pay back his creditor on time so far. The last instalment of Rs 5000 (approximately $61) is due today. The moneylender called to inform him that he would be there in the evening to pick it up. The problem is that he does not keep that much cash handy, so he'll need to withdraw from his bank account. Being a busy day at the shop, Ajit decides not to go to the branch 7 km away but to the ATM instead, which is midway between Azizpur and the next village.

Ajit has used the ATM card on a few occasions before, but always with the assistance of someone. Although he is on his own this time, he is quite sure of himself, and he takes his time punching in the details. Unfortunately, the ATM does not eject the money. The printed receipt that emerges from the machine compounds his rising anxiety, and he is just about able to comprehend that the remaining account balance is showing as just Rs 500 (approximately $6)—but there's no cash in sight.

Ajit does the only things he can: he looks around, presses the green button a few times, and even thumps the machine on the sides, hoping to dislodge any notes that might be stuck, but nothing moves. Finally, Ajit runs out of the booth and looks for the guard, who usually sits outside. When Ajit narrates his ordeal to him and shows the receipt from the ATM, the guard answers cryptically, saying that this is most likely the outcome of a technical issue, and that Ajit needs to take it up with the bank.

Ajit is beside himself with worry. He has never defaulted on his payments, and his credibility is at stake. Above all, this is his hard-earned money. It will take him several months to save this much again. This is exactly why he didn't want to switch from the post office in the first place or try using the ATM machines.

He has no idea where to go now or who to complain to about his lost money.

Anas[1] is tech-savvy and uses digital financial solutions on a daily basis. From booking tickets online for his frequent travels to making payments online on food delivery apps and monitoring his payments on his banking app, he does it all. It was little surprise then that he started using a credible-sounding third-party service to pay his credit card bills. The benefit was that he could consolidate all his cards in one place and track their payments without worrying about missing due dates. Things went well for the first few months. Then, on one occasion, he had to make a few large payments in one go.

In Anas' case, the funds got deducted *twice* from his account, but the banks to whom he had to make the payments had not received the money. So, between him and the bank, the third party probably had the money since the money had to be somewhere. Only when he tried reaching out to the third-party service provider did Anas realize that there was no way of reaching them. There was no toll-free number, no chat option, and no customer service at all.

So even though there was a customer who was very motivated to register a complaint to retrieve his money and who had the technical knowledge and wherewithal to do so, the means just did not exist. The third party, which was popular and seemingly credible in the market, had not provided any options for their customers to reach out in times of need like this.

Eventually, Anas did get his money back. It took three anxious working days and a lot of crossing fingers. Unwilling to go through a similar experience again, Anas has sworn off the app and has gone to other financial service providers to pay his credit card bills.

---

[1] The co-author of this book.

When it comes to money, people are very cautious; they count the money carefully, sometimes even more than once, before parting with it. Electronic transactions, in contrast, are just not intuitive and do not appear to be as safe, even for people who are relatively tech-savvy. From time to time, all of us make mistakes.

We have done it. Probably many of you have too. You book a flight using an online portal only to realize later, sometimes a bit too late, that you booked it for the wrong date, or you messed up the timings—you had intended to book an evening flight at 5 p.m., not a morning flight at 5 a.m. Sometimes, you realize right after you finish booking that you made a mistake. You call the airline or the online portal, and they tell you, 'Sorry, the flight was non-refundable, and to change it would require a huge penalty.' Damn! This would not have happened if you had used a human travel agent. Most of us hate it when it happens, but we soon get over it and move on.

Now imagine someone who is not as well-off as you are. In fact, financial inclusion efforts are targeted at people who are poor and cannot afford to lose any money on simple digital transaction mistakes. This fear prevents them from using any digital transactions at all. What if I send the money to the wrong account because I typed the wrong phone number? Once I press 'send', is my money gone? Can the transaction be reversed? How? If it can be reversed, that creates other problems. What if I am a seller who receives money? If the buyer can reverse the transaction after receiving my goods and services, what protection do I have?

The digital transaction landscape is not yet robust enough to deal with these serious issues. It appears that, in general, redressal mechanisms to address transaction mistakes, frauds and scams are essential before more people feel comfortable enough to switch to digital payments and transactions. Not only does a redressal mechanism need to exist, but it also needs to be immediate, simple and ubiquitous.

The Current Situation with Redressal

Sulbha Mane is a housewife. She has a husband who rides an autorickshaw for a living and two young children, whom they are trying to educate as best they can. Her daily routine of running the small household, sending her children off to school and cooking and cleaning for the family takes up most of her time. She is aware that her financial literacy is limited, but she does not let that stand in the way of her financial prudence.

Her savings are mostly in the form of small amounts of cash. It's not only easier for her to squirrel away small sums from the household budget, but when there is a sudden emergency, this money is always at hand.

There's another reason why Sulbha prefers to save on her own: the few times she's visited the bank, she felt she wasn't treated with respect. Her lack of education and her inability to read documents and sign forms would make the bank staff impatient with her. The only occasion on which Sulbha now agrees to go to the bank is when her husband goes with her. He can at least sign his name, and Sulbha feels that they are taken more seriously with him around.

During the COVID-19 pandemic, the government credited money into the accounts of eligible beneficiaries through the Pradhan Mantri Garib Kalyan Yojana. Sulbha heard about this from her friends in the shanty and mustered the courage to go to the bank and check her balance only because some of the women were going together. When they approached the counter, a bank employee informed them that it was a particularly busy day at the bank and that the women could simply dial the number on the poster and go through the IVR to check their balances instead of bothering the staff. When one of the women told the staffer that they did not know how to use the IVR, the response given was to 'ask the guard'.

After this experience, Sulbha stopped going to the bank altogether. One of the women in her group complained that this happens to them often. It's best if they send their husbands to take care of the banking work. So, that night, Sulbha, too, asks her husband to take her passbook with him the next time he goes.

What if Sulbha knew her rights as a customer and insisted that she be taken seriously, no matter how small her request was? What if the women were aware that, in the case of poor services, they could (and should) complain to the manager of the branch? Would the knowledge have been enough? Assuming that Sulbha wanted to *and* knew how to raise a complaint, she would have the option of first complaining to the bank itself about the staff's behaviour. This would require Sulbha to have the courage to ask for the manager, the confidence to walk into their office, and the ability to articulate her concerns. Given that it is already intimidating, would she even be able to contemplate these, let alone execute them? Her interaction with the bank staff, therefore, becomes defining in many ways.

The other option available to Sulbha is less intimidating. Again, assuming that she is aware *and* motivated enough, she has the option of logging an online complaint. With some help, she can download the bank's app. However, this would require her to register on the app with her customer ID and bank account details and verify them through her debit card or with a one-time password.

Sulbha is lucky that her first language, which happens to be Marathi, is being offered as a language of choice by the bank. However, she still needs to find the correct place in the app to register a complaint. These apps usually ask several questions to verify and categorize the request. In some cases, details such as account number, IFSC code and names of the receiver's or sender's banks may be asked too. Would Sulbha even get to this point? Or would she give up midway, thinking that it's easier to stop using the bank altogether?

This is not to say that Sulbha cannot use her phone well. She can. In fact, she is as intrigued by the phone as she is confounded by it. She knows there's a whole world in there, from videos to movies to WhatsApp messaging to video calls. With some of her savings, she bought herself an inexpensive smartphone. She has taught herself how to watch videos on YouTube, and her daughter has helped her learn how to send voice messages instead of text messages (which she is unable to do due to her limited literacy). Some words and commands on the screen of her phone now appear familiar. Even though she can't read them, she knows what they mean. With voice messaging, video calling and entertainment on her phone, Sulbha's needs are fully met. But despite the journey she's made, interacting with a banking app, particularly to log a complaint, is still a tall order for her. So, while she has a smartphone that can support banking and other financial apps, Sulbha is unable to use them. They require basic literacy skills that she doesn't have. Sulbha's digital and financial literacy exclude her from the process of seeking redressal through digital apps.

One might suggest that Sulbha make use of the voice-based IVR system to log her complaints if she is unable to use the banking app. This would require Sulbha to be aware of the bank's contact number to dial it in the first place and then jump through the complex IVR steps, often requiring her to authenticate with debit card details, a process that many of us find very frustrating. Due to the requirements and complexity of the IVR channels, it is not a viable option for people in this segment to seek grievance redressal. Furthermore, a survey conducted by Gram Vaani and the National Institute of Public Finance and Policy (NIPFP)[2] shows that one in five people did not know how to complain, and over 30 per cent reported they did not seek redressal for their issues due to previous bad experiences dealing with rude bank officials.

While Sulbha may lack digital and financial literacy, she at least has the use of her phone. Manjula, one of the women from the group who had gone to the branch to access her balance, does not know how to use a phone. In fact, she doesn't even own one. A decade or so earlier, when everyone around her started to carry phones, she wanted one too, so she asked her husband if they could get her one. He'd asked her why she needed a phone for herself when she could use his. Manjula tried telling him that he was barely home for her to use his phone, but the implication of her husband's words was clear: Why did she need to have conversations with anyone that her husband was unaware of? Manjula had let the matter go.

All the big decisions in Manjula's house are made by her husband. From decisions on her children's education to matters of finances and decisions related to the land they had in the village, all of it is decided by him. Having or not having a phone is not very different from those other decisions.

So, if she ever needs to talk to someone, she asks her son to call from his phone. If a relative or friend wants to speak with her, they call her son's number. If she is lucky and he is home, she gets to talk to them. Whenever there is a need for her to share her number for official reasons, like at the bank or on government documents, she gives her husband's number, and he speaks on her behalf. For Manjula, this lack of digital

[2] Katepallewar, Rohan S., and Vani Viswanathan, 'The Grievance Redressal Process for Banks Excludes Many Indians', India Development Review, March 15, 2022, https://idronline.org/article/advocacy-government/the-grievance-redressal-process-for-banks-excludes-many-indians/, accessed on July 16, 2023.

empowerment means not having to visit the bank alone or log a complaint digitally, as she does not even have access to a mobile phone to access the app and contemplate using it.

SMS 'Unhappy'[3]

In 2010, a brilliant, simple and effective consumer complaint programme was designed and implemented in the Indian state of Andhra Pradesh by Shiva Kumar, chief general manager at the Hyderabad Local Head Office of the State Bank of India (SBI). It is called 'SMS Unhappy'. Any SBI customer, anywhere in Andhra Pradesh, who is 'unhappy' about any service can send, 24×7, a simple text message 'UNHAPPY' using a mobile phone to a number 8008 20 20 20 advertised on billboards and in all SBI branch locations.

The cost of sending the text message (a few paise) is the only one incurred by the customer. The message is received and recorded in a web-based system at Shiva Kumar's office, assigned a unique ID, and a return text message acknowledging the complaint along with the unique complaint number is sent to the mobile phone used to send the message (in India, all incoming calls and text messages are free for the person receiving the call or the text message). Specially trained staff, sitting in an aptly named 'Happy Room', call the customer back to take down the details of the complaint and enter them into the system. The SBI branch manager is notified about the complaint via an automatically sent alert text message. The complaint remains open in the system until it is resolved, after which a closure message is sent to the customer.

---

[3] Batra, Rishtee, Kumar Piyush, 'State Bank of India: SMS Unhappy', Indian School of Business Case No. ISB001, Indian School of Business, February 2013, Harvard Business Publishing https://hbsp.harvard.edu/product/ISB001-PDF-ENG, accessed on July 16, 2023,

Notice minimal effort on the part of the customer: all he or she needs is a mobile phone, and even if the customer is not literate, it is easy to ask someone how to type the letters UNHAPPY to send the first text. The entire system involves absolutely no paperwork. The most brilliant part is that a permanent electronic record of the complaint is created, which can be accessed and used for monitoring, feedback and incentives. The public sector banks in India have been notorious for their inefficiency, maddening paperwork requirements and appalling lack of accountability because there is usually no simple way to complain, and even when one does complain, access is only to a local supervisor who has no incentives to follow up on customer complaints. The system implemented by Shiva Kumar appeared to work because there is now an awareness among the bank employees and branch managers that someone higher up in the system may be watching, and a record of inefficiencies and complaints could be accessed with a few simple keystrokes. The impact of the SBI Unhappy pilot could be seen in the survey conducted by MARS on customer satisfaction in several cities across India just a few months after the launch of the pilot, and the findings of the survey showed SBI to be number one in terms of customer satisfaction among all banks only in Hyderabad.

We can design a robust redressal mechanism, taking inspiration from this brilliant programme. Instead of asking customers to send an SMS, they could be asked to simply make a 'missed call' to a single, easy-to-remember number, which will reduce the cost to the customer even further to zero. Of course, the key to designing a mechanism like this is that it should work quickly and consistently over long periods of time. While the SMS Unhappy programme showed great promise and success initially, as it scaled up, its reliability has deteriorated significantly over the years.

RBI Integrated Ombudsman Scheme

The Reserve Bank of India (RBI) launched the Integrated Ombudsman Scheme in late 2021 in an effort to provide a more convenient and efficient way for consumers to address grievances related to banks, non-banking financial companies (NBFCs), and digital transaction companies. Prior to the implementation of the Integrated Ombudsman Scheme, consumers may have had to navigate multiple channels in order to file a complaint or track the status of their case. The Integrated Ombudsman Scheme aims to simplify this process by offering a single portal, email address, and physical location where consumers can report their complaints and submit supporting documents.

While the Integrated Ombudsman Scheme provides a centralized way for consumers to address their grievances, it is important to note that it is not the first point of recourse and would seem daunting to most users. Consumers are required to try to resolve their issues directly with the service provider before escalating their complaints to the Integrated Ombudsman Scheme. However, if a resolution cannot be reached through the service provider, the Integrated Ombudsman Scheme serves as a last resort where consumers can take their complaints. The two-tiered system ensures that consumers have multiple options for addressing their grievances and that the Integrated Ombudsman Scheme is only utilized as a last resort.

Despite the existence of the Integrated Ombudsman Scheme, accessing its mechanisms for resolving issues or complaints is not a straightforward process. This is particularly true for marginalized and rural communities, which often have low awareness and usage of these resources. According to the Reserve Bank of India's annual report on all Ombudsman schemes,[4] less than 5 per cent of all complaints to the Integrated Ombudsman Scheme came from rural areas, with the majority being filed online. This suggests that those with stronger digital skills are more likely to utilize these services. In addition to social factors, such as the limited involvement of women in financial matters, there are also practical barriers to seeking resolution. Some people may be fearful of negative consequences for complaining or simply find the process too confusing, lengthy, or frustrating. Basic digital literacy and financial knowledge are often required to navigate the forms and technical language involved, and even having a smartphone and basic phone skills can be necessary for those aware of their options.

In short, though redressal mechanisms exist, they are neither simple nor expeditious enough to generate enough trust to encourage people to switch to digital services.

One Nation, One Number

Any recommendations to improve the redressal process as well as the redressal experience for the complainants must accommodate different groups: the savvier urban population as well as the BOP group, which is our focus in terms of financial inclusion. A solution that is technology-heavy, with little or no human interaction, may be more conducive to urban populations, but these tend to exclude BOP groups. The inferior experiences they have after they've made an effort to participate in the financial system can set them further back than where they started.

---

[4]'Reserve Bank of India—Publications', n.d. https://www.rbi.org.in/Scripts/PublicationsView.aspx?id=21624#ES, accessed July 16, 2023.

One of our key recommendations is the establishment of a centralized grievance redressal system. What this means, in effect, is the presence of one single toll-free number and/or one single app that people with grievances can approach for a resolution, irrespective of the bank in which they hold accounts or the financial service provider through which they have availed the service.

Instead of being a body made up of bank representatives, this could be a body that is independent in its functioning but would work closely with the financial service providers to complete the redressal process. As a centralized body, it would have access to all information relating to the number of complaints, the type of complaints, the time it takes to resolve them and so on, and this data could be closely monitored by the regulator, the RBI, to ensure consumer protection. Furthermore, in the interest of transparency and to build trust in the system and the financial service providers, some data could also be made public on a real-time basis. The cost of running this centralized grievance redressal body could be shared by all the financial service providers.

For instance, Sulbha found herself intimidated by the bank staffer's behaviour towards her and could not muster the courage to ask for the manager. But with a centralized grievance redressal system, Sulbha could, with much less apprehension, call a third party and register her complaint with them.

The central system would note down the details, along with the name of the bank and branch location where Sulbha faced issues. They would assure her of a response within a pre-defined time frame. Then they would send an alert to the bank with Sulbha's complaint. Because the public and the regulator monitor all complaints and the time it takes to resolve them, the bank would be committed to fixing the issue and reporting its resolution to the central redressal body.

Sulbha gets to avoid the discomfort of a confrontation, yet she gets a solution. It might even salvage her dwindling faith in the banking system.

An added benefit of a centralized redressal system is the ease with which multi-party issues can be resolved. A common issue, for instance, is when customers withdraw cash from another bank's ATM and there's a discrepancy: the cash is not dispensed, but the amount gets debited. For many customers, this is not an easy problem to resolve. They don't know which bank to call. Should they call just their own bank or also the bank whose ATM they went to? Where do they get details about the other bank? Should they visit the other bank's branch? Do they need to visit their own branch? What if they keep passing the buck?

Instead, if they made just one call to the centralized system and the third-party body alerted both banks, the customer would get a resolution without having to worry about so many details. Both banks are obligated to respond within timelines, resolve the matter and report it back to the third-party redressal body.

Since the public and the regulator track progress regularly, neither the third-party body nor the banks can take the complaints lightly. There will be a timely closure of all complaints logged.

Digital Didis and BCs in Redressal

A second recommendation is more for the benefit of the BOP group or financially marginalized groups in rural settings. These are typically the groups that may not have the financial or digital literacy to undertake the complaint process on their own. A more 'hum-tech' approach is what is recommended for them.

In the previous chapter, we spoke about creating a network of Digital Didis. This network of Digital Didis, along with the network of BCs, could act as an intermediary between customers and the banks. Grievances such as number updates, SMS alerts not coming in on time and consecutive debits for a single transaction can be daunting for the BOP group to address on their own. With the help of an example, we shall now attempt to show how Digital Didis, and BCs could work to solve the grievances of the people at the BOP.

Sandesh Singh, who lives in the same village as Ajit, is a farmer. In fact, his house is a few lanes away from Ajit's, and he also buys groceries from Ajit's store. Last month, Sandesh was approached by a new fertilizer supplier from a village nearby who was offering better rates. After some thought, Sandesh decided to give it a try with a small order of Rs 1000 (approximately $13).

Since the supplier was in another village, he asked Sandesh to transfer the funds through UPI. Sandesh has limited digital literacy, so he took his neighbour's help to make the transfer. Since it was his first time making such a transfer, he ended up making a mistake while entering the figures. Instead of transferring Rs 1000, he added an extra zero and ended up transferring Rs 10,000 (approximately $122) instead. The amount, of course, was immediately deducted from his account, and he was left with barely Rs 200 (approximately $2). The neighbour, whose help he had taken, had no idea how to reverse the transaction. Sandesh tried calling the supplier, but his phone was switched off. Convinced that he had been cheated, Sandesh gave up.

Later that night, when he was at Ajit's shop, Sandesh shared his story with him. Ajit suggested that Sandesh speak with Sundar, the local BC. The next day, Sandesh, who had never reached out to a BC earlier, met with Sundar and described how he'd been cheated by the fertilizer supplier. Sundar assured him that the supplier had no intention of cheating him because the mistake had happened at Sandesh's end and there could be some other reason why he was not answering his phone. Sandesh requested that Sunder log a complaint and try to retrieve the money because, intent aside, the supplier is still not traceable.

Two days went by. Sandesh had still not received his money or heard from Sundar. That night, he visited Ajit again and updated him on the lack of progress. Ajit informed him that if one BC does not get the job done, Sandesh is free to go to another. They reached out to another BC, Gautam. Sandesh has no choice, so the next morning he met with Gautam and, once again, shared his ordeal.

Unlike Sundar, Gautam reached out to the supplier by sending him a chat request in the UPI app itself. In one hour, he was able to get an alternate number for the supplier and contact him. The supplier appeared to be genuinely sorry about the mix-up. His daughter had taken the phone and gone to a relative's house for a few days. She was unaware of the situation with the excess payment made by Sandesh, but when Gautam messaged her, she replied with an alternate number to call her father on. The supplier promised to immediately transfer the remaining Rs 9000 (approximately $110) back. Sandesh is relieved, and he thanked Gautam profusely.

An interesting feature of the Digital Didi and BC networks is that there is competition involved. Each time a Didi or a BC manages to close a task and get a redressal, they are paid for it. Also, customers from the BOP can choose who they go to. If one of the Didis or BCs is unable to get them redressal, as in Sandesh's case, they are free to ask another for help.

This keeps the network competitive. Because each redressal is monetized for them, it ensures that they have the incentive to follow up and get closure for the customers.

Insurance Against Mistakes

It is now common to see QR code-based payments accepted in a variety of places, including small local stores or kiranas. UPI in India has experienced a significant surge in both the number of transactions and the volume of those transactions. While demonetization may have initially sparked this trend, it was the COVID-19 pandemic and the associated concerns about handling physical currency that truly cemented the position of UPI in the Indian financial landscape. The widespread adoption of digital payment methods, including UPI, has made it easier and more convenient for consumers to make purchases and has also helped reduce the reliance on physical cash.

While these figures accurately reflect the current state of digital transactions in India, they may not paint a complete picture, particularly when it comes to rural areas. In these regions, digital transactions are less frequent and tend to involve smaller amounts. There are several socio–economic factors that contribute to this. One of the primary causes is internet connectivity, which can be unreliable in areas with low population density. Fewer mobile towers serve these areas, leading to more frequent issues with connectivity, network drops and the slow resolution of problems. This can cause consumers to lose trust in digital solutions. Traditional habits and patterns, such as the preference for cash payments, can also be difficult to change. Many rural shopkeepers, traders and merchants may be unfamiliar with receiving digital payments and may view them as 'credit'. Even the payment confirmation via SMS may not be sufficient for sellers who have limited literacy, especially in English, which is the preferred language of many digital players. The dispute resolution process also lacks credibility among the rural population, further undermining trust in digital solutions. There's also the issue of cyber scams that deter customers from going digital.

Rahman Shaikh is a kirana owner in Anjar, Gujarat. After much persuasion from his children, he finally acquired a QR code for the account linked to his shop. It's still early days for him, and he is watchful and wary of the new technology. So far, everything seems to be okay. Fortunately for him, a majority of his regular customers still prefer cash, just like him. It's the youngsters who choose to pay with their phones, even for amounts as little as Rs 12 (approximately $0.15). When that happens, he checks their phone screens for the animated tick mark that pops up when the payment is confirmed. However, he makes a mental note to tally the amounts later.

A few weeks later, he is sitting at his regular tea stall and chatting with a few friends. The conversation turns to new-age methods of payment, and one of the men in the group, Bala, informs them of a scam that the youngsters are pulling off. To Rahman's dismay, he listens to how some customers buy goods worth a significant amount and then, instead of paying on the shopkeeper's number, they pay their friends or family members the money (to be retrieved later). When the payment confirmation shows up on the screen in the form of an animated tick mark, they quickly flip the phone over and show it to the unsuspecting shopkeeper, who thinks that he is the beneficiary of the payment. Before the shopkeeper can verify anything, the customers flee on their bikes.

Rahman is so taken aback that he decides to revert to cash-only transactions. When his children try to explain later that there are ways to check immediately if he has, in fact, received the payment or not, he gets upset with them and tells them to come and sit in the shop with him if they want him to retain the QR code.

Rahman is one of many small merchants and traders who are simply wary of moving to digital platforms because of limited or no awareness. Rahman's primary source of information is the grapevine at the tea stall, where facts are embellished with rumours, stories and even personal analyses.

What, then, are the alternatives available? How can India ensure a digital system with minimal pitfalls, pervasive trust and robust redressal? Bank Balance Batao (BBB) is a helpful precursor to most systemic changes that can be implemented. Being able to check their account balance immediately after completing a transaction would alert consumers on whether or not the money transfer has taken place.

However, the voice notification service alone may not be sufficient to address the needs of India's financially underserved populations. In order to truly improve financial inclusion, there must be mechanisms in place to quickly resolve transaction grievances as soon as they occur. It is essential to have simple and efficient systems to resolve disputes and errors in a timely manner to facilitate greater financial inclusion.

Our recommendation here is to introduce insurance for transaction mistakes. Banks could engage a third-party insurance provider who would basically underwrite the losses of faulty transactions, and it could help streamline the redressal process even further. For instance, think of how air travel is insured at the time of flight booking. At a minimal cost, one can insure themselves against loss of baggage, flight delays, etc.

Similarly, all digital transactions could be brought under the scope of an insurance provider. As part of the claims process, the provider would immediately be able to identify disputed transactions, transactions made in error or any other kind of grievance that customers may have. The comfort of knowing that an insurance provider is underwriting potentially faulty transactions can boost the confidence of BOP customers and would most likely encourage more digital use.

The insurance provider investigates each case and provides redressal on behalf of the bank. However, unlike insured air travel, where delays and baggage losses are well documented, in the case of digital transactions, errors may be more difficult to track. In other words, fraudulent claims may rise once the customer base is aware that their transactions are insured. In such cases, the insurance provider could analyse any patterns or trends that emerge and inform the bank. Say, for example, that there's a sudden spike in disputed cases from a certain digital platform, from a region or city, or even from the same number. The banks can be alerted, a more thorough investigation can be conducted, and the fraudulent claims can be denied. Major credit card companies employ a fraud management system along these lines. E-commerce giant Amazon also absorbs the costs of fraudulent product returns and has an alert system that investigates any spikes in returns from suspicious customers or locations.

One may argue that this insurance would be an additional cost to the financial service providers, for whom profit margins on digital payments are slim to none. However, one must see the alternative scenario where, due to a lack of digital payments, there is an increased dependence on cash, particularly with the BOP customer base, which often receives direct-benefit transfers from the government and is often withdrawn completely in cash, and there is a considerable cost to financial service providers in making these cash withdrawals possible in all parts of the country. We propose that these cost savings be factored into the pricing of such insurance systems that could be employed by financial service providers.

Eventually, engaging an insurance provider could be a win-win for all three parties concerned: the customers, particularly those at the BOP, which we are trying to include in the financial system; the banks and financial institutions and the insurance providers, who are able to generate a new line of business for themselves.

Chapter Seven

Overdraft[1]

---

[1] This chapter is based our op-ed which we had co-authored with Lisa Nestor and was published in the *Economic Times* on 20 November 2020.

The most pernicious effect of being excluded from financial services is not the fact that the poor are unable to earn adequate returns on any savings they may have. Yes, the poor do save, even if their savings are small and in the form of cash, gold, cattle or other small assets.[2] But often, these savings are not enough to take care of unexpected emergencies: an illness, a death, an accident, a loss or the breakdown of a critical asset such as a vehicle. The first option at such times is seeking financial assistance from family and friends.

However, that is often not available quickly enough or reliably. Reaching out to family and friends could also generate shame, loss of face and dignity, and uncomfortable feelings. What is needed in such emergencies is a simple, flexible and quick loan at reasonable interest rates.

Expensive Loans Are Profitable, Hence Ubiquitous

Loans are available. They are even quick, simple and flexible. But alas, they are not cheap. Local moneylenders, pawn shops, payday lenders and check-cashing loan outfits are prevalent all over the world.

When Anil Mahato, a farmer in Panderia, Jharkhand, found his wife Roopa collapsed on the small patch of land they farmed together, he thought it was fatigue. Days later, she was just as ill and not getting any better. The worse she got, the

---

[2]Agrawal Niyati, Ponnathpur Rakshith, Seetharaman Sahana, and Misha Sharma, 'Insights from the All-India Debt and Investment Survey 2019', Dvara Research, February 1, 2022, https://www.dvara.com/research/wp-content/uploads/2022/02/AIDIS-Slide-Deck.pdf, accessed on July 16, 2023.

more difficult it became for Anil to care for her—he had to take on all her chores too. He was tending to the small plot that gave them vegetables to sell in the local haat, looking after two young children and cooking for them all by himself.

When it began to look like more than just a seasonal illness, Anil finally took Roopa to the government dispensary, where she was diagnosed with tuberculosis a few days later. What seemed like a surmountable obstacle slowly turned into a near-hopeless situation for the Mahatos.

Roopa became weaker by the day, despite the medication. Anil could only do so much on the small plot by himself. Their already paltry income diminished. The fine balance that they had achieved with their daily income from the land and their limited spending was irrevocably disturbed. Roopa's medication was expensive. The loss of her labour was an added burden. There were no savings to tide over the situation either.

Anil did not have many relatives left in the village, but he tried asking the remaining few for help. He even asked Roopa's relatives, something he had never done earlier. However, everyone's life seemed to be as precariously balanced as Anil's until a few months ago. The best they could do was offer some sympathy and amounts of money so small that he would only be able to afford two days' worth of medicines for Roopa. Feeling defeated and ashamed of having to ask for money, Anil finally decided to 'look outside'. There was nothing of value left with the Mahatos. The little jewellery they had was already pawned a few years ago to get the land they now till. The bank was not even an option for Anil. In his already depleted mental state, he did not have the resilience to jump through the hoops that a bank loan required. Besides, he needed the money immediately. As early as that same evening. And so, Anil made that long walk to the place he'd dreaded all his life: the moneylender's house.

His need was a far stronger driver than his fear of what might happen in the future.

Informal sources of credit, though expensive, are lifesavers in many situations when no other source of credit is available. In Roopa's case, they were quite literally lifesavers. With the cushion of the borrowings, Anil was able to focus better on her health and slowly nurse her back to life. In a few months, she was back on her feet and working on the land again. They had a huge loan to pay off together.

Regulators have treated such informal sources as predatory, and providers of such loans have been portrayed in popular culture as evil. Calls to ban, prohibit and outlaw such activities surface from time to time all over the world. However, such practices persist, despite the negativity and scorn faced by the providers, because they are profitable and, more importantly, because they fulfil the unmet need for credit.

Rather than joining the clamour of opposition, our view is that the only way to address this important issue is to design and provide an alternative that is just as simple, flexible and convenient, yet fair and affordable.

Cheap Loans Are Not Profitable, Hence Not Ubiquitous

In the earlier chapter on Microangels, we said that MFIs are not profitable at scale due to their high cost and are therefore not ubiquitous. Here we explain how the high operating costs translate into exorbitant interest rates for microloan borrowers and also affect the profitability of MFIs.

Let us consider two micro-entrepreneurs. Rosa has a small business idea that requires an initial investment of Rs 10,000 (approximately $122) and will generate Rs 11,000 (approximately $134) at the end of the year—a return of

10 per cent per year. Nandini also has a business idea that requires an initial investment of Rs 1000 (approximately $13) and will generate Rs 1500 (approximately $18) at the end of the year—an annual return of 50 per cent. Neither Rosa nor Nandini have any savings to finance the initial investment.

Pro-Fit is a (hypothetical) microlender that finances micro-entrepreneurs such as Rosa and Nandini. Pro-Fit incurs a cost of Rs 400 (approximately $5) each year for every borrower it serves—say, to hire a loan officer who will try to ensure that borrowers are using their loans for business purposes and not for consumption and that they will pay back their loans in a timely manner. To keep things simple, let us assume that the cost of capital for Pro-Fit is close to zero. Rosa will be charged an interest rate of at least 4 per cent on her Rs 10,000 loan, while Nandini will have to pay 40 per cent interest on her Rs 1000 loan, in order for Pro-Fit to cover Rs 400 (approximately $7) in costs.

Now, suppose there is another microlender, No-Fit, which is not as cost-efficient as Pro-Fit and has to incur a higher annual cost for each borrower of, say, Rs 600 (approximately $7). No-Fit cannot possibly make any profit by lending to borrowers like Nandini, but it can break even by lending to borrowers like Rosa if the interest it could charge is at least 6 per cent. Pro-Fit, which is in competition with No-Fit, can choose an interest rate of 6 per cent for borrowers like Rosa and an interest rate of between 40 per cent and 50 per cent for borrowers like Nandini, and that will ensure that No-Fit is unable to compete with Pro-Fit and make any profit.

Thus, Pro-Fit, in effect, becomes the sole surviving lender and is profitable by lending to both types of borrowers. At an interest rate of 6 per cent, the company makes a profit of Rs 200 (approximately $2) on a Rs 10,000 (approximately

$122) loan to Rosa, and at the 45 per cent interest rate that Pro-Fit charges Nandini, it makes a profit of Rs 50 (approximately $0.061) on a Rs 1000 (approximately $13) loan. Rosa keeps Rs 400 (approximately $5) after repaying her loan, and Nandini keeps Rs 50—which is better than nothing.

Box 2: Is Interest Rate Capping in the Interest of the Poor?

Being consistently profitable makes Pro-Fit decide to go public by doing a blockbuster IPO that gets everyone's attention (think Compartamos Banco in Mexico and SKS Microfinance in India), and a debate ensues about whether Pro-Fit should be making profits off poor people and if it should be charging interest rates as high as 45 per cent over its cost of capital.

In one scenario, imagine that the proponents of keeping interest rates low (such as Muhammad Yunus) dominate public opinion and convince the regulators to impose an interest rate cap of 10 per cent on all microloans. Pro-Fit stops lending to smaller borrowers like Nandini and continues to lend to larger borrowers like Rosa at a 6 per cent interest rate. Nandini and other small borrowers like her can no longer borrow from microlenders, and she is either driven out of business or forced to borrow from a local moneylender who charges her a very high rate of interest, leaving little or nothing for her after paying off the moneylender.

In another scenario, regulators leave the markets alone, and over time, seeing Pro-Fit make healthy profits after a successful IPO, other microlenders start to innovate and become as cost-efficient as Pro-Fit, provide healthy competition to each other, and force Pro-Fit to drop its interest rates from 45 per cent to 40 per cent.

In which of the two scenarios above are the poor better off?

Charity Is Cheap and Cost-Efficient . . .

Before you jump to an answer, consider yet another possibility that has not received sufficient attention in this debate in microfinance circles. Suppose we were to subsidize lenders like No-Fit so that they could compete and be able to offer lower interest rates to borrowers like Nandini. One may argue that the social benefits of empowering small borrowers like Nandini are large enough for a strong case to be made for such subsidies. How large should the subsidy be, and how low should the interest rates be?

We argue that the case for a zero-interest rate is the strongest. The reason is that every time an MFI charges a positive interest rate, it must hire loan officers, and for every loan worth Rs 1000 (approximately $13), it would have to spend about Rs 400 (approximately $5) every year just to be able to collect the loan back. Imagine that—a 40 per cent loss in collection costs alone—and that if the MFI did not have to spend this money on the collection, it could fund another borrower like Nandini in roughly two years.

A loan with a zero-interest rate is nothing but charity (which is what charitable organizations like the Bill and Melinda Gates Foundation prefer when they finance health and energy initiatives). Many renowned microcredit proponents, most notably Yunus, often discard charitable donations arguing that, 'Charity dollar has only one life, whereas a microloan for business has many lives.' This is misleading. A charity dollar should be seen as a zero-interest loan with a very long maturity. We must trust that the poor are capable of making judicious decisions, and there is plenty of evidence that suggests that this is the case when the money is in their hands.[3]

. . . But is Limited, Thus Cannot Be Ubiquitous

The trouble with charity and subsidies is that charity dollars are limited. However, savings is another source of funds, which like charity, does not suffer from the wasteful depletion of resources spent collecting loan repayments. With savings, as with charity, the micro-entrepreneur becomes her own financier, which is cost-efficient because it prevents large and wasteful spending on collection and monitoring. The increasing attention and emphasis on micro-savings in recent years is thus a welcome change in the landscape of microfinance and social entrepreneurship. Micro-savings provide perhaps the most effective competition for high-cost moneylenders and MFIs. Investing efforts in ensuring that we develop ideas and create institutions that allow the poor to save effectively may be worthwhile.

The evidence of the poor becoming their own financiers successfully is sketchy.[4] Dean Karlan, Sendhil Mullainathan and Benjamin N. Roth conducted experiments in India and the Philippines where market vendors were offered cash grants to avoid taking expensive daily loans from moneylenders. However,

---

[3] Banerjee, Abhijit, Esther Duflo, and Garima Sharma, 'Long-Term Effects of the Targeting the Ultra Poor Program', *American Economic Review: Insights*, 3 (4): 471–86.

[4] Ananth, Bindu, Dean Karlan, and Sendhil Mullainathan, 'Microentrepreneurs and Their Money: Three Anomalies', 2007, https://scholar.harvard.edu/files/sendhil/files/microentrepreneurs-and-their-money-three-anomalies.pdf, accessed on July 16, 2023.

they found that most vendors returned to borrowing from the moneylenders within six weeks.[5]

There are also many behavioural studies showing that the poor need help forming the habit of saving, and research by Anandi Mani, Sendhil Mullainathan, Eldar Shafir and Jiaying Zhao illustrates how poverty impedes cognitive function and weakens the will.[6]

Perhaps Yunus did have a point—a microloan with a fixed repayment schedule does have positive behavioural habit-forming benefits. But at what cost? A 25 per cent or greater per year loss in transaction costs is too high a price to pay! If only the cost of providing the loan could be substantially reduced to as close to zero as possible. We believe that a combination of mobile technology and digital identity allows us to precisely do that.

Is PMJDY's Overdraft Facility the Answer?

The Pradhan Mantri Jan Dhan Yojana (PMJDY) is a financial inclusion programme launched by the Government of India in 2014. It aims to provide access to financial services, including savings and deposit accounts, credit, insurance and pensions, to all Indian citizens, particularly those who are poor or unbanked. The goal of the programme was to offer equal financial access to everyone.

---

[5] Karlan, Dean, Sendhil Mullainathan, and Benjamin N. Roth, 'Debt Traps? Market Vendors and Moneylender Debt in India and the Philippines', American Economic Review: Insights, 2019, 1 (1): 27–42. https://doi.org/10.1257/aeri.20180030, accessed on July 16, 2023.

[6] Mani, Anandi, Sendhil Mullainathan, Eldar Shafir, and Jiaying Zhao, 'Poverty Impedes Cognitive Function', Science, 2013, 341 (6149): 976–80. https://doi.org/10.1126/science.1238041, accessed on July 17, 2023.

The PMJDY overdraft facility is a feature of the PMJDY programme that allows account holders to withdraw more money than they have in their account, up to a certain limit, by borrowing from the bank. This facility is intended to provide a safety net for PMJDY account holders who may need financial assistance in times of emergency or when they incur unforeseen expenses. However as per the information provided by the minister of finance to the parliament in 2019 the usage of this facility remains very low at less than 1 per cent of PMJDY account holders.

Pitambar Lal had migrated to a tiny suburb of Muzzafarpur about a decade ago from his village in rural Bihar. Though he had come alone to look for employment, within a few months of finding a job in a small welding unit, he moved his wife and firstborn with him. His second child, a daughter, was born in the government-run hospital in the city—something that gave great solace to Pitambar, who had nearly lost his older child, a son, during childbirth in the village.

Now, even though he was still poor, Pitambar was content with his life in his one-room shanty. The money was less, but it was regular. Unlike in the village, everything was available nearby and easily. The nearest dispensary was not an hour away, and he didn't have to beg and plead to be paid on time for the work he did in others' fields.

Last year, he overheard a conversation about the PMJDY scheme and decided that it would be a good idea to open an account for himself. He followed through, and though he had to withdraw the Rs 500 (approximately $6) he'd deposited initially, he still had the account. A few months back, around the same time that his wife was contemplating starting a small embroidery set-up from home, Pitambar found out about the overdraft facility that his PMJDY account offered.

To help his wife get started, he decided to apply for an overdraft and pay it off slowly as the income started to roll in. However, Pitambar was not fully aware of all the details.

To be eligible for the PMJDY overdraft facility, account holders must have a PMJDY account that has been operational for at least six months and have a good track record of regular deposits and timely repayment of any previous overdrafts. The overdraft limit is typically set at a maximum of Rs 10,000 (approximately $122), and the interest rate on the overdraft is around 7 per cent per annum.

Pitambar unprepared for the amount of paperwork that would be expected of him and the time it would take for him to get it done. The steps he would need to go through when applying for the overdraft facility under the PMJDY scheme is as follows:

1. He would need to find and visit a branch of his bank which could be several kilometers away from his home as it is not possible to request this facility at BC point.
2. Fill out a two-page loan application-cum-undertaking form requesting for the PMJDY overdraft facility. The form will ask for personal and financial details such as name, address, PMJDY account number, and employment details.
3. He needs to then submit the form with required documents and wait for the bank to review the application and decide whether to approve or reject it based on the scheme policies and guidelines.

The guidelines and rules of eligibility areas as follows:

- The applicant has to have a PMJDY account and have been satisfactorily operating it for a minimum of six months.

- Only one member of a particular family can avail themselves of the overdraft facility. Female members are usually preferred for this.
- The account holder should also have a good credit history, although for an overdraft of up to Rs 2000 (approximately $26), no guarantee is needed under PMJDY.
- For loans beyond Rs 2000 the following conditions are used to determine the loan amount:
  - Four times of average monthly balance
  - Or, 50 per cent of credit summations in account during the preceding 6 months
  - Or, Rs 10,000 (approximately $122) whichever is lower

In Pitambar's case, he fell short of the minimum duration by just a month. He had opened his account five months earlier, whereas the overdraft facility was available to accounts that had been operational for at least six. Unfortunately for him, he could not help his wife start her business on time. He did, however, find out that female family members were preferred, so they had two options: wait for a month and apply again from Pitambar's account or open an account for his wife and wait for six months.

Pitambar and his wife decided to apply with his account, even though it meant postponing their plans. At least they would be able to initiate the process again after a month. In a few months, Pitambar's wife might be running her own small embroidery business from home.

Yes, PMJDY's overdraft facility is cheap and useful, but it is too complicated for most borrowers, who need something far simpler and easier to understand; therefore, it is not ubiquitous.

Mukti: Cheap, Efficient and Transformational

When Laxmi's husband died suddenly a year ago, she did not even have enough savings to give him a proper cremation, observing the bare minimum rituals. In such circumstances, time is of the essence, and she was forced to borrow Rs 5000 (approximately $61) at an exorbitant interest rate from a moneylender. She was grateful for Sushma Tai, the moneylender, who was there when Laxmi needed her. She received the money right away, without any paperwork or hassle, and repayment was simple and the terms flexible, even if exorbitant. Sushma Tai charged a fixed fee of 3 per cent upfront, making her total repayment principal Rs 5150 (approximately $62) plus 3 per cent per week recurring interest—not an atypical rate charged by moneylenders to poor borrowers in need.

Laxmi could have used the overdraft facility to borrow the Rs 5000 (approximately $61) she needed, available with the PMJDY account that she was persuaded to open a few years ago when there was a big drive encouraged by Prime Minister Narendra Modi. Alas, she did not qualify. To be eligible, the account holder had to maintain a positive balance for several months, perform many transactions, fill out an application, etc. Laxmi never knew about the eligibility criteria and in any case, they seemed too complicated to her.

In January 2019, the then finance minister Piyush Goyal, replying to a Rajya Sabha question, stated that less than 2 per cent of PMJDY account holders successfully filled out the paperwork and met the eligibility criteria that allowed them to avail of the overdraft facility.

What we suggest is a solution called 'Mukti', named after the Hindi word, that means freedom or liberation. What if the overdraft facility in Laxmi's PMJDY account had allowed her to withdraw Rs 5000 of 'Mukti Money' instantly even if her account had a zero balance and she had not done many transactions over the past few years—which is indeed the case with a substantial fraction of PMJDY account holders? In addition to providing immediate access to emergency cash in times of crisis, the repayments could also be structured so that they are simple and easy to understand. Furthermore, this facility could be used not only to collect repayment but also to inculcate the habit of saving—a cushion of savings Laxmi may need again in the future on a rainy day. This model would then have the best feature of Sushma Tai's moneylending model, which is immediacy, and it could satisfy the Yunus model's apprehensions about long-term self-sufficiency.

Here is a repayment programme that is easy to explain. Essentially, Laxmi is told that she can avail of an interest-free loan of Rs 5000; in other words, the bank gives her Mukti Money. She has to return the money in fifty equal instalments of Rs 100 (approximately $1) each. But she must also save an additional Rs 100 per week for fifty weeks over the year, and she can withdraw the saved Rs 5000 (approximately $61) at the end of the year, a week after she has also repaid her loan completely. The programme could go so far as to waive or postpone two out of fifty payments for unforeseen circumstances.

With this programme, Laxmi is receiving a Rs 5000 loan and repaying it with fifty equal instalments of Rs 100, which amounts to an interest-free loan. But with her Rs 100 per week savings deposits, which will accumulate to Rs 5000 after fifty deposits, she is also providing a systematic interest-free loan to the bank. So, the bank more or less breaks even if it levies just a small fee of Rs 100.[7]

Notice, however, that not only does Laxmi get critical financial help for her emergency, but in one year, she manages to save an additional Rs 5000 as well. Once Laxmi reaches a savings balance of Rs 5000, she will be allowed to avail herself of the overdraft facility again. Now she is doubly protected.

Compare this to what her situation would be with the loan from Sushma Tai. An equivalent weekly repayment schedule of fifty equal payments over the year to what she paid the moneylender is approximately Rs 200 (approximately $2) per week, and at the end of the year, she will have zero savings. Moreover, when she borrows from the moneylender and repays, she does not build any verifiable credit history. With her overdraft and savings history, Laxmi is building an important financial record of borrowing and saving, which will help her get more financial services in the future. This is the essence of financial inclusion.

There is, of course, the obvious concern that some borrowers would withdraw the money from the overdraft facility and not repay any of it. The cost of such a default strategy is that the borrower will only be able to do it once. Given that the PMJDY is linked to Aadhaar, which has many other benefits associated with it, most borrowers are not likely to jeopardize

---

[7]The loan that the bank provides occurs during the first six months, and the loan that the borrower provides to the bank occurs during the second six months of the year. A simple net present value calculation would show that the bank would need to charge a small fee to break even. Adding a small default premium for the possibility of default by some borrowers would require a higher fee. Our estimates suggest that a fee of Rs 100 is sufficient (which amounts to a return of 8 per cent per annum on an average loan size of Rs 2500 [average from 0 to Rs 5000] for a period of six months) to compensate the bank for both the time value (approximately 3 per cent annual) and the default premium [of approximately 5 per cent per annum], which is reasonable for a microfinance loan).

the flow of all such future benefits for a one-time gain of Rs 5000 (approximately $61).

Simplifying the overdraft facility in the way we suggest would make it **Simple, Cheap** and **Ubiquitous**. Thus, it will not only prevent poor people from falling into the clutches of moneylenders, but it will also teach them the habit of saving after they are confronted with a cash crunch in an unexpected emergency. The Mahato family's brush with unexpected illness and Laxmi's pain at the sudden death of her husband are both deeply unfortunate incidents. However, both of these incidents are also opportunities to teach them lessons in financial prudence with Mukti in a way that no financial literacy programme can ever come close to.

Chapter Eight

Insurance

The Buddha spoke about the four inescapable truths of life: birth, old age, sickness and death. Leaving aside the inevitability of mortality, sickness and death are inextricably linked to the vagaries of nature. When it's not nature, then accidents and man-made calamities pose a threat to life and general well-being. In essence, no human life is immune to ill health, accidents, misfortune or financial losses. No matter what the adversity, be it health, death, accidents, unemployment or the failure of the monsoons, all of these eventually result in a financial crisis; such is the interconnectedness of human life and money. Another peculiarity of this reality is that no one really knows when terrible times may come. And that's why, as a collective, we've come up with mechanisms to safeguard ourselves against crises.

What Are these Mechanisms?

Short answer: community. You take care of me when I need help. I will take care of you when you need help. That requires two things. Groups of people. And an ability to write social and formal contracts (collective intelligence and the ability to remember and reciprocate). Hence, social institutions and social insurance.

But what if the entire group is threatened? That requires you to enlarge the group for diversification. But a larger group means less reliance on informal mechanisms like reciprocity. We need formal mechanisms called contracts, which lead to individualism and reliance on market insurance. Formal development of these systems takes resources, which poor societies can't afford. They are compelled to rely on informal mechanisms. But there may be a good hybrid as societies move from informal to formal. That is what we are suggesting in this chapter. Informal mechanisms at small scales and formal mechanisms at large scales.

Insurance is one such mechanism. While largely formalized, some forms of insurance have existed informally in the past too. The joint family system that continues to thrive in India has the need to mitigate financial risk at its very core. There are other sociological factors too, but one cannot ignore the economics of the arrangement. These forms of informal hedging come with several limitations. The ability of a single big shock to wipe out entire families or even whole communities is just one of those limitations.

Ratnesh Bhai Shah is a pottery and terracotta craftsman who, along with his newlywed wife, was travelling to Mumbai when a massive earthquake hit Bhuj, Gujarat, in 2001. Back home in the village of Chobari, his extended family of four brothers, their wives, and several children were caught unaware on that fateful morning of 26 January. It took Ratnesh Bhai several sleepless days and nights to finally get home, only to find that his entire family had perished and that the place he had called home his whole life had been levelled by the quake. Ratnesh Bhai Shah's family had been in the pottery trade for generations, and by choice, they had continued to live and work as a unit. This shared model of living and working had saved them many a rainy day. He still remembers his grandmother telling them tales under the night sky about how the elder Shah and his brothers had managed to complete a festive order on time only because they were together. And how the goddess at the temple had blessed them with prosperity for generations after that. One of the reasons why the cousins and brothers never ventured far from the family home was this. The rest was also sheer economics. When they worked and lived together, expenses were divided, and incomes multiplied.

After the earthquake, there was nothing left. Neither the home nor its inhabitants. Along with the family members that Ratnesh Bhai Shah lost, he also cremated all the security, shield and surety he'd ever had. The Shah family's assiduously cultivated method of staying together to minimize risks tumbled in front of the massive earthquake.

While on the subject of the collective need to hedge risks, it is perhaps noteworthy to talk about Joseph Henrich and what he postulates about collective intelligence.

Joseph Henrich is an anthropologist and evolutionary psychologist who has studied the concept of collective intelligence. He defines it as the ability of a group to solve problems and make decisions that are at least as good as those of the average individual in the group. This is because the collective intelligence of a group can be greater than the sum of its individual members' intelligence.

In his book, *The Secret of Our Success: How Culture is Driving Human Evolution, Domesticating Our Species, and Making Us Smarter*, Henrich discusses how human societies have evolved to become more intelligent and cooperative over time. He argues that this has been driven by the development of cultural institutions and processes, such as language and institutions for economic exchange, which allow individuals to coordinate their actions and work together to achieve common goals.

Henrich's theory is that human societies have been able to solve problems and make decisions more effectively than other animals because of their ability to coordinate their actions and work together to achieve common goals. This has allowed us to develop complex cultural institutions and technologies that have given us a competitive advantage in the natural world.

For example, Henrich points to the development of language as a key factor in human evolution. Language allows us to communicate complex ideas and share information with one another, which has enabled us to learn from each other and work together in ways that other animals cannot. This has allowed us to develop complex social institutions, such as economies, governments and armies, that have given us a competitive advantage over other species.

This very urge to 'come together' to form collectives is the bedrock of our solution around insurance for the BOP population.

Let's build upon the irrefutable fact that crises occur. From minor mishaps to major catastrophes, we deal with troubles on a daily basis. Crops fail, earthquakes level entire cities and unheard-of viruses bring the whole world to a grinding halt. We know that society and formal insurance allow you to share risks among people.

Prasanta Bora is a fisherman from a small village near Pahukata, Assam. He has always relied on the river for his livelihood, and his fishing boat is his most valuable possession. But a couple of years ago, things took a turn for the worse. A severe flood swept through the northeast, devastating homes and businesses and leaving many people struggling to rebuild their lives.

Prasanta's fishing boat was one of the casualties of the flood. The powerful waters had completely destroyed it, leaving him with no means of earning a living. Prasanta had always known the importance of insurance, but he had never been able to afford it. As he stood amid the ruins of his livelihood, he couldn't help but regret his decision to go without coverage.

Without insurance to fall back on, Prasanta was faced with a difficult choice: either find a way to rebuild his business or give up on what he knew best and loved most: fishing on his boat. Given how devastating the flood was, Prasanta had few options to explore outside his line of work and decided to do everything in his power to rebuild his boat. He worked tirelessly, pouring all of his savings and energy into the project. But despite his efforts, the costs of rebuilding were simply too high. For a daily earner with limited savings, there was only so much he could do.

As the weeks passed, Prasanta's financial struggles only grew. He was unable to pay his bills or provide for his family, and the stress of the situation began to take a toll on his health.

In the end, Prasanta had to borrow money from a relative who lived in another state just to be able to get back on his feet. With their support, he was able to rebuild his boat and reclaim his livelihood. Prasanta is still paying off the loan and will continue to do so for the next few years. But now he knows firsthand the importance of being prepared for the unexpected and has vowed never to go without coverage again.

What if we took the best features of formal and informal solutions and integrated them to create a third, new way of making insurance a better and more robust tool? And what if we could do this without changing anything fundamental about insurance itself?

The solution we propose is that formal insurance should not be targeted at the individual but at the community or the whole group. At the group level, informal mechanisms work better, even though they may not be enough to cover very large risks.

Sharing Risks

The origins of formal insurance mechanisms can be traced back to the early days of human civilization. In ancient societies, people would come together and form groups or associations to pool their resources and provide mutual aid and protection against various risks and hazards. Over time, these informal arrangements evolved into more organized and formal systems of insurance, in which individuals would pay premiums to a central entity in exchange for financial protection against specific risks or losses.

The first known insurance contract dates back to the Code of Hammurabi in Babylonia, which outlined the rules and regulations governing trade and commerce, including provisions for compensating merchants for losses due to theft or damage to their goods. Later, in ancient Greece and Rome, formal insurance arrangements were developed to provide protection against the risks associated with shipping and maritime trade.

In contemporary times, the development of formal insurance mechanisms has been closely linked with the growth of the global economy and the increasing complexity and interdependence of various industries and sectors. Today, insurance is a vital part of the financial system, providing individuals and businesses with a means of mitigating risks and protecting against potential losses.

Despite the slow but sure evolution of formalized insurance, informal insurance mechanisms have also continued to persist. These refer to informal arrangements or systems that provide protection against risks or losses, often in the absence of formal insurance contracts or institutions. They are often found in communities or societies where access to formal insurance is limited or unavailable.

Informal insurance mechanisms can take many forms, depending on the context and the needs of the individuals or groups involved. For example, in some cases, informal insurance may involve individuals or households coming together and pooling their resources to provide mutual aid and support in the event of a loss or disaster. In other cases, informal insurance may involve the use of social networks or community ties to provide protection against risks or to share resources and expertise.

Informal insurance mechanisms can be effective in providing some level of protection against risks, but they often lack the stability, reliability and legal enforceability of formal insurance arrangements. In addition, informal insurance mechanisms can be vulnerable to fraud, abuse or other challenges, making them less effective or sustainable in the long term.

Limitations notwithstanding, people have traditionally relied on informal mechanisms to protect themselves against the vicissitudes of nature. These informal mechanisms usually take the form of saving for a rainy day; transfers from family members less affected by adverse outcomes; charitable donations by more fortunate members of society and government relief aid in various forms (e.g., waiver of debt payments, free electricity, water and food for affected people, etc.). At the level of an entire country or a state, relief may come from foreign aid and charity mobilized by relief organizations across the world. Even though such informal risk-sharing mechanisms are important in civil society, they bring their own uncertainty, are not very reliable and are generally limited to a local area—international aid and support arrive only for extremely visible calamities. As a society, we ought to be able to do better than that.

Raadhi Bisa is a mother of three who lives in a small hamlet on the edge of a forest in tribal Odisha. Each day, one of her chores is to trek into the depths of the forest and collect firewood. After her husband leaves for the fields and her children have gone to attend school in the next village, Raadhi calls out to some of the other women who live close by, and they all start their daily trips to the forest.

A month earlier, Raadhi began to feel short of breath but didn't say anything to anyone. This continued for a few weeks until Raadhi could not keep pace with the other women anymore. Every now and then, she would stop and rest. Getting up and walking again was as difficult. One day, when she had severe chest pains along with very short breaths, she finally confided in her husband at night. Mani, her husband, was troubled and, the very next day, took Raadhi to the local dispensary in the same village where their children went to school.

The attendant gave Raadhi a few colourful pills and wrote something on a sheet of paper. These were tests that Raadhi needed, but the single-room medical hut was not capable of doing them. They would have to go to the government hospital, which was further away. Though the government-run hospitals are free, Mani knew they would need money for other incidentals. There was the travel itself and the usual list of things to buy when the test itself was free. And there was no telling what the medicine bill would come to.

Neither Mani nor Raadhi were formally insured. In their small hamlet, everyone depended on informal methods of insurance, such as pooling resources with neighbours and relying on the generosity of the community. Mani, too, turned to the local community for help, as they had done in the past for smaller medical expenses.

Their generous donations were sufficient to cover the cost of the initial visits and tests, but when it came to treating Raadhi's diagnosed heart condition, the entire hamlet's collective might also did not suffice. The informal insurance methods that had served them well in the past were simply not enough to handle a critical illness.

As the days passed and Raadhi's condition worsened, Mani began to lose hope. He had always been strong and self-sufficient, but now he was helpless and at the mercy of the healthcare system. He recalled the day when some 'sarkari'-looking people had come to their little village and offered to give them something called insurance. All they would have to do is provide details to fill out a form and pay a token cost that was called a premium. The money was little; even Mani could afford it, but the general suspicion and mistrust of strangers who appeared only when they needed something from the villagers made most of the village stay away from the documents. There were also too many cumbersome questions to answer about pre-existing conditions and other health details. No one in the hamlet knew much about these things. Now, sitting outside the ward in the government hospital where Raadhi lay in bed, weak and miserable, Mani wondered if he should have just gone ahead with the paperwork. Perhaps they would've paid enough to treat Raadhi in a better hospital?

Finally, after weeks of struggle, Mani received some good news. A charitable organization had heard about Raadhi's plight and offered to cover the cost of her medical bills. They would move her to a bigger, privately run hospital in Sambalpur and take care of every last rupee needed, even the cost of moving her there.

Raadhi and Mani's story ended well because of a stroke of luck. Not all uninsured BOP people are as lucky. Charity and philanthropy are matters of chance, and one cannot plan one's life around the availability of these. What Raadhi needed was comprehensive health coverage to diversify her risks, which she did not have.

What Could I Explore that Would Lead Me to the Diversification of Risk?

This is what the financial markets allow us to do. In the jargon of financial economics, many risks can be diversified. For example, one can protect oneself against many risks by buying insurance. Many insurance products, such as life insurance, automobile insurance and health insurance, exist and are used by millions of people. But as Robert J. Shiller, Nobel Laureate in Economics, has pointed out, there are many other risks we face in life—for example, unemployment risk—and the level of formal insurance, or livelihood insurance, is woefully inadequate. Furthermore, in a research paper by Shawn Cole, Xavier Giné, Jeremy Tobacman, Petia Topalova, Robert Townsend and James Vickery,[1] they show, using evidence from India, that there are many barriers that have severely limited the use of formal insurance contracts. They suggest that many poor people don't buy insurance because they face liquidity and credit constraints, do not trust formal insurance products, and sometimes the advantages of insurance, such as large payouts in adverse situations, are not often visible. Innovations in the private sector that address these barriers can increase the use of formal insurance products.

While these solutions may work and certainly ought to be given a shot, we suggest a different solution altogether. We propose a marriage of formal insurance contracts with informal insurance mechanisms. Large institutions, such as self-help groups, local banks and cooperatives, temple churches, mosques and gurudwaras, even state and central governments, can buy formal insurance against the risks faced by people in the communities they serve. Risks such as drought, floods, earthquakes, disease, negative shocks to local livelihoods, etc. The local institutions can then use their informal mechanisms—shelters, food langars and cash disbursements— to help the people adversely affected in their area. They could

---

[1]Cole, Shawn, Xavier Giné, Jeremy Tobacman, Petia Topalova, Robert Townsend, and James Vickery, 'Barriers to Household Risk Management: Evidence from India', 2013, American Economic Journal: Applied Economics, 5(1): 104–35.

even buy group insurance for their members' protection against critical risks, which would be a lot cheaper than if individuals were to buy it on their own. This eliminates the need for (or supplements) the uncertain and unreliable aid that may or may not materialize when needed.

The paper by Cole et al. suggested that the poor are very reluctant to make upfront payments for formal insurance products; they are so cash-constrained that they are willing to ignore the fact that if they are faced with adversity against which they have not bought insurance, their situation would be very dire. A reluctance to pay upfront cash premiums is even prevalent in middle-class families, who can, in fact, afford to buy insurance products.

On the other hand, people are often willing to help others when they are able to by making charitable donations (recall the success of PM Cares Fund during the pandemic) and offerings to gods in temples, masjids, churches and gurudwaras. What we suggest is that these institutions, when they receive such donations, should leverage the funds to buy formal insurance products for their communities, making their protection even stronger. In other words, Mr Smith of St Joseph's Parish in Burnpur will be covered by his church insurance when he acquires a critical illness that he can't pay for, Malathi Nair from Panoor can claim insurance from her temple when her crops fail due to floods, and Bhaktawar could seek help from his Gurudwara when a fire destroys his harvest. And none of these would come with the loss of dignity that accompanies alms meted out or charity handed over; these would be legitimate claims settled through a formal insurance process. One could argue that when they were seeing better days, each of the beneficiaries might even have contributed to the fund that enabled the religious institutions to procure the coverage in the first place.

Markets work well at larger scales and from a distance. Informal mechanisms work well at smaller scales and locally, where information and trust problems are resolved more effectively using social networks. An appropriate combination of the two would provide more robust protection against life's risks.

Chapter Nine

Credit Where Credit Is Due

Santosh Mathur, thirty-five, lives in a two-bedroom apartment in a suburb of Mumbai with his wife and six-year-old daughter. Santosh works for an MNC, while his wife has a job at an IT firm. Their daughter attends a school about 6 km away from home. Between the two of them, they make sufficient money to live a comfortable life that can afford them full-time domestic help and a vacation every year. They are also able to occasionally support their parents, who still live in their hometown of Prayagraj.

Once a month, usually at the beginning, Santosh and Meeta, along with their daughter, go to a nearby mall to shop for the entire month. They buy groceries in large quantities, often picking up things that they don't really need. But this is a better bet than having to make multiple trips. They barely have any time during the week anyway. Invariably, these grocery runs to the mall end with lunch at the food court and a few other activities that give them all the feeling of a day out.

Brijmohan Jaiswal is also a thirty-five-year-old who lives in a single-room house in a slum not far from the Mathurs' apartment building in Mumbai. His village, in fact, is just forty-five minutes away from where Santosh Mathur's parents live in Prayagraj. Brijmohan has been riding an autorickshaw in Mumbai for the last ten years.

It is Brijmohan who drops Santosh's daughter off at school every day. He knows the Mathurs well, and they trust him. Sometimes, when Santosh and his family need a ride, they call him directly, and he obliges. For example, on their monthly trip to the mall. Recently, app-based cabs have started to take a lot of time to arrive, and very often, they charge arbitrarily too. Besides, when they call Brijmohan, he helps them lug the heavy grocery bags up to their third-floor apartment.

Brijmohan has a wife and two school going children, and they all live in the same cramped space. Brijmohan's earnings come in on a daily basis, and they barely have any savings. His wife, Manorama, uses his daily income as judiciously as she can to ensure that they have the necessities that they need. She buys just enough groceries and vegetables to last them a few days. For one, they don't have a refrigerator or a large enough kitchen to stock too much at home. Besides, they usually have just enough cash to buy for a few days at a time.

Now that you know both of our protagonists, let's understand their spending patterns:

When the Mathurs go to their regular supermarket in the mall, Meeta Mathur prefers to buy a 5-kg dal packet for Rs 590 (approximately $7) that would last them for five weeks. But when Manorama goes to her usual kirana, she buys 500 g dal (often loose and unbranded) for Rs 65 (approximately $0.79). In fact, she might do this twice a week, or ten times over, by the time the Mathurs run out of dal.

Manorama does this because she only has a few hundred rupees to buy groceries for a few days. If she did have Rs 500 (approximately $6) to begin with, she would've probably put it away for a rainy day, not knowing when her husband's daily income might be affected.

Brijmohan makes about Rs 600–800 (approximately $13–$10) a day after paying for the rent of his autorickshaw and the fuel. Some money is put aside for rent, some is sent back home, and some of it goes towards the children's education. There have been enough days when Brijmohan has come back with less than Rs 500 (approximately $6). There are days he loses time in the queues at the CNG station. There are days when he has to have repairs made. There are days when he just doesn't manage enough business. And that is why Manorama manages the remaining money from the household budget as frugally as she does.

Coming back to the two women's spending habits on groceries, essentially, Manorama is paying Rs 650 (approximately $8) for the same quantity of dal that Meeta is paying Rs 590 (approximately $7) for. Arguably, Manorama is also buying unbranded, inferior-quality dal. The paradox here is that Meeta pays less because she's richer, and Manorama pays more because she's poor. This phenomenon is also called the poverty premium or the poverty tax.

The concept of the poverty premium is relevant to many countries, including India. In India, low-income individuals and households may face a number of additional costs that can make it more difficult for them to make ends meet. These costs can take many forms, such as higher prices for goods and services, higher fees for basic services, and higher interest rates on loans.

Besides Manorama's example, another example of the poverty premium in India is the cost of healthcare. Low-income individuals in India may have limited access to preventive care, leading to higher costs for the treatment of preventable conditions. They may also be more likely to rely on emergency rooms for medical care, which can be more expensive than primary care. In addition, the cost of prescription drugs can be a significant burden for low-income individuals, as they may be unable to afford the cost of necessary medications.

Another example of the poverty premium in India is the cost of transportation. Low-income individuals in India may be more likely to live in areas with limited public transportation options, leading to higher costs for transportation to work or to access other necessary services. In addition, the cost of owning and maintaining a vehicle can be a significant burden for low-income households.

The poverty premium is a complex issue influenced by a variety of factors, including access to credit, bargaining power and the costs associated with serving low-income markets. It has been an important area of research in the field of development economics, which focuses on issues of economic growth and poverty reduction in developing countries.

One of the early researchers on the poverty premium was economist Amartya Sen, who has written extensively on the issues of poverty and inequality. In his work, Sen argues that poverty is not just a lack of income but also a lack of access to resources and opportunities that are necessary for a minimally decent life. Sen's work on the concept of 'capability deprivation' has been influential in shaping our understanding of the ways in which economic systems can disproportionately impact the poor.

In an earlier chapter, we discussed how moneylenders charge the poor exorbitant rates of interest because they do not have access to formal channels of borrowing. We had also spoken about how moneylenders are villainized despite providing a much-needed service. (Imagine what the poor woman or man would do if, after feeling excluded by the financial systems, they did not even have a moneylender to go to.) Well, here's a situation that does not involve moneylenders but still ensures that the poor pay a very high premium for just being poor.

In the case of Manorama Jaiswal and the dal she frequently buys at Rs 65 (approximately $0.79), it is equivalent to offering her a loan of Rs 590 (approximately $7) upfront but with an instalment plan in which she makes ten equal payments of Rs 65 (approximately $0.79). That amounts to a loan at an effective annual interest rate of over 200 per cent! Again, no greedy moneylenders are present here, just your friendly neighbourhood kirana guy. And this picture repeatedly occurs all over the country, for hundreds of millions of the poor population. If you find the presence of moneylenders charging 40 per cent or 50 per cent annual rates of interest disturbing, you should be equally concerned about the exorbitant costs that poor people are forced to incur for their everyday purchases because we have not offered them the simple and inexpensive financial credit products that you and I get easily. This is where financial exclusion hurts the poor the most.

Regrettably, this travesty is not limited to basic necessities. Think of what happens when the Mathurs need to buy a new washing machine. Despite having some cash handy, they may decide to keep it for a rainy day because they have multiple other options to choose from. They can just buy the washing machine on credit. Banks, and even retailers themselves, are eager to offer you credit with a plan to repay the principal and interest using a convenient repayment plan that may involve Equal Periodic Instalments (EPI) over a long but fixed time period.

All they have to do is stroll into a store during one of their mall visits, and several salespeople will approach them with options that allow them to buy whatever they need in simple EPIs. A key decision variable when Santosh and Meeta decide whether or not to accept the EMI plan is the effective interest rate they are charged. Sometimes, retailers would offer much lower interest rates, or even no interest, often termed a no-cost EMI plan, to induce them to buy the product sooner than they would if they were to save up the entire cost of the product before choosing to buy it with cash. Lured by the no-interest option, the Mathurs, in fact, made a spot decision to add a microwave oven to their cart and round off the EMI.

Now, what if the Mathurs were offered an EPI plan with an effective annual interest rate of over 100 per cent, just like the one that the Jaiswals are forced to pay for basic necessities? They'd probably scream 'Preposterous!' and refuse the offer on the spot, accusing the retailers of fleecing them and even posting a tweet or two online. The Jaiswals, however, have regularly, routinely and ubiquitously accepted these 'preposterous' rates month after month for years.

Fortunately, there is a simple fix for this giant problem. It does not require any regulatory intervention or policy reform. It only requires that the poor feel secure about digital transactions, allowing easy and low-cost access to digital financial services for everyone.

The owner of the kirana where Manorama shops is a middle-aged man known in the neighbourhood as Pandey, which is his last name. Pandey could offer a 5-kg dal package at a low-cost weekly instalment plan if the weekly repayments were simple, with a near-zero transaction cost and tied to a credible customer identity that made default by the customer a highly unattractive option. This can be accomplished by offering automatic credit for those with bank accounts, with automatic authorized deductions each week. The chances that people will default and jeopardize their future access to these and many other benefits for a one-time gain of a few hundred rupees will be negligible and can be easily priced into the product cost.

According to a World Bank report, the poor spend around 56 per cent of their income on food and basic necessities. A back-of-the-envelope calculation shows that by buying small quantities, the poor end up effectively paying an 8 per cent tax on their income (or a 16 per cent tax on food expenditure). From this, we estimate that the poor in India lose around Rs 88,000 crore (approximately $10.7 billion) each year from their income on EPI interest payments—which probably exceeds what the poor pay in exorbitant interest charges to moneylenders!

One simple way to implement this idea is to convert the current kiranas into franchises of large FMCG companies. When Manorama goes to the kirana to ask for a 500-g packet of dal for Rs 65 (approximately $0.79), the owner offers her a 5-kg packet instead for Rs 590 (approximately $7) and suggests that she pay Rs 60 (approximately $0.73) now and the remaining nine payments in Rs 60 weekly instalments for the next nine weeks. If she is digitally savvy, she can make these payments online, take the help of a Digital Didi, or come to the store and make the payments in person (the last option can be discouraged so that the kirana owner's time and indeed Manorama's time are not wasted in going to the store just to make the payments). The possibility of Manorama defaulting will be thwarted because if she fails to make the payment, this facility of bulk buying will be stopped for all future purchases. Similar to the system envisaged in the Microangels chapter, a default on any one kirana will amount to a default on all kiranas that are part of the same franchise.

All the pieces required to make it happen are already here. We just need to put them together with some ingenuity and creativity. And speed.

Chapter Ten

Cash 'n' Ladder

We have argued that cash savings at home are a poor investment because they produce no interest and can be stolen or squandered on temptations. Most importantly, cash lying idle at home does not contribute to building a financial history that can act as intangible collateral that could provide access to credit that may be needed for emergencies, education and entrepreneurship. Despite varying literature and intuitive wisdom, several groups, including the elderly and those in need of care, people with limited digital literacy, etc., continue to keep money at home.

Savings instruments do provide some return but are quite often not sufficiently attractive unless one is ready to put them away in fixed deposits (which make them largely inaccessible). In recent times, there has been an increased interest in attracting investments in equity mutual funds from the lower middle-income population. If these instruments have the potential to change the game for the middle-income population, one then wonders if the poor should also have easy access to equity investments through mutual funds.

Well, not yet.

Equity investments come with inherent risks, and it is important to approach them with caution. It can be tempting to be swayed by the prospect of high past returns; and the poor at the BOP are especially vulnerable to this kind of temptation that hints at disproportionate returns. This fascination with monetary fates turning overnight is already seeded into our culture through folklore and stories and is fuelled by the more recent bourgeoning of reality television and game shows that can turn people into 'crorepatis' overnight. It's not surprising, then, that for a large section of the population at the BOP, the possibility of putting in some money and watching it multiply overnight is quite believable.

What holds true in general about 'past performance not being a guarantee of future successes' is perhaps truer in the case of mutual funds. Though all mutual fund companies do inform potential customers about this, perhaps the voice of warning is slightly subdued in comparison to the articulate and sticky catchphrases that proclaim, 'Mutual Funds Are Right' and 'Mutual Funds Are Good'. These lines, especially when spoken by trusted and liked celebrities from sports and cinema, have the ability to stay longer in the minds of the audience than a prosaic warning that says 'past performance is not a guarantee for future success'.

In addition, investing in mutual funds can be complicated, which may make it difficult for people with lower income or limited financial knowledge to participate. Unfortunately, fraudulent schemes sprout every now and then, particularly targeted at people from the BOP as well as lower middle-income groups, with promises of spectacular returns that ultimately result in financial loss. It's essential to be vigilant before committing and subsequently losing funds to them.

Ponzi Pawns

Debjit Acharya worked as an accountant in Kolkata in the early 2000s. On his very limited income, he was raising three children and taking care of an invalid older brother. Back home in eastern West Bengal, his extended family was fighting over the ancestral land holding. Debjit, the only one who had moved to a city for work, was secretly hoping that the land would be divided so he could take his share and invest it for the future of his children. He had no interest in agriculture, and their holding was so remote that there was no foreseeable chance of selling it at a premium.

To his relief, that's exactly what happened. The family decided to sell the land and divide the money. Debjit received Rs 4 lakh (approximately $4877), a princely amount for him. He kept this in his bank for some time, trying to figure out the best way to make it grow. For a lower middle class man with three children, of whom two were girls, his financial responsibilities were immense.

It was during this time that he first heard of the Sarada group. It was a conversation he chanced upon at his workplace, where some senior men in suits were discussing the explosive growth in portfolios that Sarada investors were gaining. Sunil was a man of numbers, and he also had Rs 4 lakh in his account that needed a home to grow. He mustered the courage and spoke to one of the men later in the day. To his surprise, the man was very welcoming to Debjit and shared all the information about the group and how he could put his money in.

Convinced that a scheme with such eloquent and well-heeled proponents could not go wrong, Debjit put all of his money into Sarada.

Not too far away, in the mangroves of the Sundarbans, Minoti and her family of five survived on whatever the river could give them. Minoti was enterprising, a little too much so for her mother, who believed in accepting what fate had given them.

By the age of twenty-two, Minoti had saved Rs 1 lakh (approximately $1219) on her own by working odd jobs since she was fifteen. Every summer, she went to the city and lived with a relative while she worked and saved every penny she could. It was during one such visit that she heard about the Sarada scheme from a neighbour of her relative, whom she travelled with every day on the bus. The returns promised were unheard of, and at first, Minoti couldn't believe that any of it was true. However, from what she gathered from the neighbour, she was receiving her returns—she even showed her passbook to Minoti.

Minoti did not wait for the next year to return to the city. She came two months later, her mind and purse made up. She took the neighbour's help and put her entire life's savings into the scheme. In fact, Minoti was even offered the role of an agent—she could take the scheme back to the Sundarbans, and for each new investor, she'd be paid a commission that Minoti could only dream of. A whole summer in the city could not make her that kind of money.

For the next six months, Minoti travelled from one hamlet to another in the Sunderbans delta. She started with the people she knew. Though people were poor, the scheme was inclusive—it accepted as little as Rs 100 (approximately \$1). The glossy marketing material and Minoti's own testimony of receiving astronomical payouts lured dozens of them to put their hard-earned savings into the scheme.

Minoti was beginning to make real money, and her mother, too, was on the verge of coming around. Back in Kolkata, Debjit couldn't believe his luck at overhearing the conversation that fateful morning. His financial plans had accelerated, and now he was dreaming of a real life where his children would go to engineering colleges. Perhaps he would even be able to buy a house in the city.

All this fell apart like a stack of cards in a span of months. When the Sarada scam came to light, the fall was as catastrophic as it was swift. The principal amount that was bringing in the returns, was fully and completely gone. Minoti had a backlog of commissions that she had no doubt would be released. But neither the unreleased commissions nor her principal materialized. Debjit had been getting returns, but the sum of these was nowhere close to what he had put in.

In effect, Sarada, which was projected as a scheme to help turn the poor from the BOP into investors, was actually stealing from those very poor people. And the poor, in their undying optimism and hope of multiplying money (which no amount of hard work seemed to be able to do), lost everything they had.

Sarada may be one of the most malicious schemes yet, but Ponzi schemes are by no means strangers to the country. Sahara is another large-scale enterprise that had a dubious end—for its owners as well as for the people who invested in it. The IMA Gold fraud from 2019 in Bengaluru is another example of exactly how the poor are lured in with the promise of unprecedented returns and then are subsequently paid some high figures to consolidate trust before the principal is snatched away entirely from the poor.

All that Glitters

The most common investment, besides savings and fixed deposits that people in India make is an investment in gold. No matter how badly off a family may be in terms of money, they most likely have some gold between them. In fact, selling gold is usually the last resort for a family in troubled times. The mere mention of 'selling gold' has a connotation of the worst of times and being absolutely at the end of one's means. Once a family in India sells its gold, it usually has nothing left.

Given the nature of gold as an ornamental metal, it's mostly kept in the form of jewellery. In any case, wealth is passed on from parents to their daughters in the form of jewellery. This is why grandparents and parents start to buy gold for their children, particularly daughters, because it will come in handy later.

Murali Reddy bought a 1-gram gold coin the day his daughter was born. It was his gift to her. Since then, it has become a ritual for him. Every year, on birthdays, festivals and special occasions, he buys her a gold coin. This way, he reckons, he'll be putting aside Rs 20,000–30,000 (approximately $243–$365) a year in gold. And it will be in his daughter's name. It's not that she has any particular use for the coins because she can't wear them. For that, there are small gifts from relatives and grandparents. Or if Murali's wife, Lata, manages to save some money and get something small made for their daughter.

In fact, this is the precise way in which Lata's parents saved for her marriage—in gold. Even when there was little money in the house, there was always gold. Money could get spent, savings consumed, and investments could turn out to be dubious, but gold? Gold would remain as shiny and comforting as ever. No matter what, you have your gold.

This is why so many people save in gold—because of the popular belief that gold never decreases in value. And the goldsmith will not shy away from reminding you of this when you appear to be in two minds about spending those few extra thousand rupees on a larger piece than the one you came to buy.

However, the truth is that gold, despite popular perception, does not always increase in price. For example, the average price of gold per 10 g in India in 1991 was Rs 9873 (approximately $120). By 2022, it had gone up to Rs 52,985 (approximately $646). That looks good, doesn't it? But while the price in 2000 was Rs 14,973 (approximately $183), it went down to Rs 10,362 (approximately $126) in 2006. It went up to Rs 31,902 (approximately $389) in 2013, but again went down to Rs 24,854 (approximately $304) in 2015.[1]

---

[1] The data used for the price of gold per 10 g in India from 1990 to 2022 was sourced from the RBI, according to an AI app Chatsonic.

The volatility of returns on gold is quite high, and the average returns over long periods of time could be no better or worse than investing in safe government securities. For example, if you had invested Rs 9873 (approximately $120) in government securities in 1991 in India, it would have become Rs 52,980 (approximately $646) by 2022—but without the roller-coaster price rises and falls associated with gold.[2] Gold is best thought of as a hedge against market risk. That is, the price of gold tends to move oppositely from the overall portfolio of stocks. So, having a little bit of gold in addition to a portfolio of risky stocks makes immense sense.

But investing in gold alone is not a great investment.

So, should the poor stay satisfied with the meagre returns that savings instruments offer while the middle- and upper-income populations get richer as the economy grows? We suggest a simple alternative.

Cash 'n' Ladder

The sage advice about risky equity investments is that you should only invest an amount that you can afford to lose. That is what risk-bearing capacity means. We suggest a simple product that we call *Cash 'n' Ladder*.

Cash 'n' Ladder offers a floor on how low your investment can fall. For example, one can promise that the principal amount you invest will be preserved. Any interest that is accumulated on the principal amount could be invested in a risky stock portfolio.

For instance, let's suppose you invest Rs 100 (approximately $1). Let us also suppose that interest on a one-year fixed deposit is 8 per cent per annum. One can invest Rs 92.60 (approximately $1) in a fixed deposit, which will grow to Rs 100 in a year. This

---

[2] According to Chatsonic, the data source for investment in Indian government securities is the RBI.

guarantees that in one year, your principal investment of Rs 100 (approximately $1) is guaranteed. The remaining Rs 7.40 (approximately $0.090) can then be invested in a risky stock portfolio. This guarantees that the minimum you will have at the end of the year is Rs 100.

In addition, imagine that a Rs 7.40 investment in stocks grows to Rs 12 (approximately 0.15 which is not guaranteed). Your balance at the end of the year is now Rs 112 (approximately $1.5). You can now be asked again if you want to put a higher floor on your investment. You are climbing a ladder, so to speak. You can decide that the new ladder step is Rs 110 (approximately $1.5). That can now be guaranteed by a fixed-deposit investment of Rs 101.85 (approximately $1.25), which at 8 per cent will grow to Rs 110. The remaining (Rs 112 – Rs 101.85) Rs 10.15 can again be invested in a risky stock portfolio. And so on.

This ensures that your savings grow on a ladder on which you never go down but only go up. Sometimes in big steps, sometimes in small steps, and in the worst case, you stay at the same step on the ladder.

It is time we included the poor in our growth journey. We can do it responsibly, with complete transparency, and with the safety that the poor so value. Cash 'n' Ladder is simple to implement; it can be easily explained by, say, Digital Didis, and it can also be offered by regulated financial institutions, making it a ubiquitous product. Cash 'n' Ladder will then pass the SHUb test as well.

Chapter Eleven

Goal-Linked Savings

Chunni Devi lives with her husband, two children and three goats in Himmatpur, a small village outside the city limits of Jaipur, the capital of Rajasthan. It should be an ordinary day like any other for Chunni. But her resolve today is distinct in its lack of ordinariness. She gave the finishing touches to a plan last night as she lay awake on the mud floor of her home while the others slept: She will start a small stitching business from home. Nothing grand, just a basic hemming and darning set-up. She knows there's a professional tailor in the village, but he's often overworked and late in returning orders. There'll be enough work for her if she starts, and there will be grateful customers too.

Chunni Devi even calculated her initial investment cost while lying awake the previous night. A sewing machine will cost her Rs 12,000 (approximately $146). If she can squirrel away Rs 1000 (approximately $13) a month, she will have adequate funds exactly a year later, when her daughter will be ready for secondary school. Putting aside Rs 1000 every month is a stretch, but she's thought that through as well. The Chaudharys, who live five houses away, have needed an extra pair of hands since their daughter-in-law gave birth. Chunni will offer them her services. She knows they will gladly accept.

Thus invigorated, Chunni Devi goes about her morning chores with a hint of a smile that mystifies her family.

Not so far away, in Dholapur, a village bigger than Chunni Devi's, Devraj works as a field officer with a public sector bank. Devraj's bank had led a very successful drive under the PMJDY, where all unbanked people in the surrounding villages were given a zero-balance account with them. He was even acknowledged as a top performer by the regional head of the bank for his role in this endeavour. It's been several years since the initiative, but many accounts are still at either zero balance or have deposits lower than Rs 5000 (approximately $61).

According to the research conducted by our team at ISB, the bank would lose money on these accounts because the annual cost of maintaining a zero-balance account is Rs 120 (approximately $1.5). In order to generate Rs 120 in revenue so that they break even, the bank requires the customer to either have deposits of at least Rs 5000 or take a loan of Rs 5000.

Over the last few months, the branch manager, Rajni, has been facing considerable pressure from the regional head to reduce the number of loss-making accounts at the branch. At their weekly review meeting, she informed Devraj and the other field officers that they must reduce these accounts. They are all aware of the reality: it is not possible to give out loans to most of the account holders because they lack proof of income. Increasing the minimum balance for savings accounts is not an option either because the account holders just don't have the money to spare. For the few who do have the sum of Rs 5000, the interest earned on the deposits is trivial when compared to the time and effort required to make cash deposits. And so, left with little choice, Rajni has announced that, in addition to reducing the existing numbers of these non-profitable accounts, they will not be opening any new zero-balance accounts unless a government target compels them to.

Back in Himmatpur, it's been six months since Chunni Devi first thought of her plan. With some difficulty, she has managed to put aside exactly Rs 1000 (approximately $13) every month. As is the case with many others like her at the BOP, Chunni Devi too has a bank account at Rajni's branch, but it is only used to receive government transfers. Her savings lie now, in a bundle of 500-rupee currency notes, at the bottom of a small tin that contains a few of her valuables as well. Even though her family knows about this, they have better sense than to use it for anything else. The tin and its contents are sacrosanct. In this regard, Chunni considers herself fortunate. She knows of enough families in her village itself where the woman tried to save money the same way, only to find one night that the husband had stolen it to get drunk or the son had stolen it to spend it in the city. But not Chunni's family.

Each month, when she puts the notes in, she feels a surge of enthusiasm. She's getting closer to starting her business; one less month to go.

While Chunni Devi has been saving assiduously and Devraj has been thinking of ways to meet the branch goal for his bank, another small plot is playing out not too far away. Around 50 km to the southwest, in the suburbs of Rajwada, Natwar Singh has just been hired as a field sales executive for Vazir Enterprises, a chain of stores dealing in electronics and FMCG. Natwar reaches work that morning, filled with the kind of zeal one usually has on the first day. Like the other sales executives, he, too, has had a geographical area marked out for him; this will be his terrain. All homes, localities, villages and communities in this area are his to cover. Armed with a list of shop names, phone numbers of potential customers and flyers with details of discounts and offers, Natwar leaves the office to start his sales campaign. He's determined to outshine his peers.

For a whole month, Natwar works hard. He changes buses, walks on foot, takes shared autorickshaws and does whatever else it takes to reach people and groups, housing colonies and even villages. Despite the rejections he faces on a daily basis (some people just don't need the things he's selling; others prefer to go to an actual shop), Natwar, fuelled by his ambition to beat his peers, has managed to rack up a fairly decent figure for the first month. He is just three days and Rs 10,000 (approximately $122) away from meeting his first sales target. A single sale is all he needs to achieve his goal.

In reality, the individual stories of Chunni Devi, Devraj and Natwar never really converge. They continue to struggle in their separate lives within a 50-km radius, completely unaware of each other's presence and unaware that all of them together could solve the others' problems.

So, here's an alternate scenario for us to consider:

Natwar and the company he works for, Vazir Enterprises, could look for people like Chunni (who are planning to buy at a later date, as opposed to looking for customers who are ready to buy *now)* and offer them an incentive to save formally, the incentive being a discount on their products. Vazir Enterprises is willing to offer this discount because it increases their sales to people who would not have saved without the incentive. So now, not just Chunni Devi but a dozen other families in the village also have the incentive to save by getting a product for less.

There's another scenario that is an offshoot of the one above:

If Chunni Devi decides to save monthly at the bank because Vazir Enterprises has incentivized it, the bank now has good, usable data on her savings habit. This information can be used to give her a loan, thus allowing Vazir to bring the sales forward.

Here's how the scenario might play out: At the end of six months, Chunni Devi is offered a bank loan for the remaining Rs 6000 (approximately $73) so that she can fund the purchase of her sewing machine. Vazir Enterprises absorbs the interest component, which makes the loan to Chunni Devi effectively interest-free for her. She continues to deposit her Rs 1000 (approximately $13) into the bank every month for the next six months, as she had planned anyway, except she now owns a sewing machine six months earlier than she originally hoped to.

What if instead of saving her money in hard cash at home, Chunni Devi had gone to a bank, started saving with them and told them expressly about her goal of buying a sewing machine? What if, instead of spending hours of his time and the company's money, Natwar could partner with a bank? So, for example, if the cost of marketing activities and discounts that Vazir Enterprises spend on acquiring a customer amount to Rs 100 (approximately $1), then Rs 20 (approximately $0.24) from this could be spent on providing incentives to the bank in exchange of making customers like Chunni aware of Vazir Enterprises and the discounts offered by them. The remaining Rs 80 (approximately $1) can be passed on as benefits and discounts to the customers directly, either by way of interest payments on EMIs or direct discounts.

This solution is unique in its ability to benefit each of the parties involved. Here's what they get:

Our protagonist, **Chunni Devi,** gets a bank account to start with. She also earns a nominal interest on the money that she deposits every month. Though not an incentive, it's still money at the end of the day. Since her savings are goal-linked and the bank knows of this, they make a seller available to her. She gets to buy a sewing machine well before she would have otherwise acquired one. Her earnings start rolling in much sooner, too.

Natwar, on behalf of Vazir Enterprises, he makes a sale to Chunni Devi and to dozens more like her who have goal-linked savings with the bank. The Vazirs are willing to absorb the interest component of Chunni Devi's bank loan as well as pay extra to the bank for informing Chunni about them and the discounts they offer. Despite the costs, the whole operation is still profitable for them because of the volume of sales they make.

Bank—Rajini's branch and the bank are able to reduce the percentage of accounts that are not generating any revenue as the customers now have the added advantage of discounts on top of the regular interest offered on savings. The bank is also able to exploit the regular savings data and offer credit to customers without any income documents, thereby adding an additional source of revenue to these accounts from the BOP.

An important part of this whole discussion remains while the alternatives are all focused on making Chunni Devi and others like her buy things, they are not designed to lure them into a consumerist trap. On the contrary, since the idea is to encourage goal-linked savings, people at the BOP will be encouraged to save only for things that they really want or need. It's difficult to imagine Chunni Devi squirrelling away her hard-earned Rs 1000 (approximately $13) every month to buy something like a television or any other gadget for which she has little use. When savings are goal-linked, they may actually go so far as to foster thrift and build prudence in the person choosing to save in such a fashion. In fact, even in this scenario, Chunni Devi enjoys the freedom to change her mind right up until the point of purchase. She may decide to change her goals or may decide to retain the money for other purposes, such as an emergency in the family, for example. It is, after all, Chunni Devi's money and is quite safe in the bank.

Chapter Twelve

Ayushman Bharat Plus

We have argued that the lack of insurance is probably one of the most terrifying exclusions that the poor face, especially during the times of an unexpected illness, death or calamity. In an earlier chapter, we also proposed a hybrid solution marrying informal insurance with formal insurance for a population as large as the one in India. In this chapter, we suggest how to fortify formal insurance schemes such as Ayushman Bharat.

What is Ayushman Bharat?

It is one of the largest ever nationwide insurance schemes launched by any government in the world. It was launched by the Central Government in 2018 with the aim of providing coverage to over 100 million poor and vulnerable families, or approximately 500 million beneficiaries. The staggering scope of the programme offers coverage of up to Rs 5 lakh (about $6130) per family per year for hospitalization related to secondary and tertiary care. This includes a wide range of medical treatments, such as surgeries, medical procedures, and diagnostic tests.

The Ayushman Bharat programme replaces the Rashtriya Swasthya Bima Yojana (RSBY) and the Senior Citizen Health Insurance Scheme (SCHIS) programmes, both of which were previously in place to provide health coverage to certain segments of the population. By combining these schemes, Ayushman Bharat aims to provide a more comprehensive and accessible health insurance solution for the identified target population.

Since Ayushman Bharat is portable across the country, beneficiaries can access care at any public or private hospital that is empanelled under the programme. This allows for greater flexibility in seeking treatment and ensures that beneficiaries are not limited to a specific geographic area. In theory, this allows beneficiaries to seek out specialized treatments in other locations as well as opt for treatment in locations where they have better non-medical support, like family or social support.

The programme is entitlement-based, which means that eligibility for coverage is determined using the Socio–Economic and Caste Census (SECC) database. This database is used to identify families that are economically disadvantaged and may not have the means to pay for necessary medical treatment on their own. By providing coverage to these BOP groups, Ayushman Bharat aims to reduce the financial burden of healthcare and make it more accessible to a larger segment of the population.

To control costs and ensure that hospitals are not overcharging for services, payments for treatment under Ayushman Bharat are made on a package rate basis, allowing the government to determine the cost of a specific medical treatment in advance.

The National Institution for Transforming India (NITI Aayog) has developed an IT platform to facilitate the programme's smooth operations. This platform is designed to be modular, scalable and interoperable, and is to be used to facilitate paperless, cashless transactions between hospitals, beneficiaries and the government. The platform is expected to improve efficiency and reduce the administrative burden associated with managing the programme.

Prabha Rani worked as a domestic helper in New Delhi in March 2021. The previous year's lockdowns had completely depleted her and her husband Mukesh financially, so at the beginning of the year, they made their most difficult decision yet. They left their two young children with Prabha's sister in their village near Prayagraj and left for work in different states. Prabha took up a full-time domestic help job in Delhi, while Mukesh started work at a construction site in Goa. When the second wave of COVID-19 infections hit and rumours were rife, Prabha's first worry was not for herself but for her children. She spoke with Mukesh, and it was decided that they would both immediately head home by whichever means were possible. Between their pooled savings, they were hopeful of lasting for at least a month or so. After that, they would have to look for employment all over again. At the time, it was important for them to be home. Prabha's sister would not be able to hold the fort for too long in a situation like this.

However, it wasn't easy for either of them to reach home. Prabha's employers refused to let her leave because they were worried about her returning with infections. Mukesh was told that he would be denied his money altogether if he left. Somehow, they both managed to leave with whatever little they could manage. Prabha arrived three days later, and Mukesh arrived in a week.

Unfortunately for them, Mukesh contracted the virus on the way home. Although he recovered, he had inadvertently passed it on to Prabha, whose condition started to worsen by the day. Mukesh rushed her from hospital to hospital, but to no avail. There were just no beds available.

The Ayushman Bharat scheme, under which they are both covered, relies on private healthcare providers for service delivery. When the government hospitals turned them away, a desperate Mukesh started his rounds of private hospitals. However, they, too, turned Prabha away because all their beds were occupied, most likely by patients who could pay. Most network hospitals have very limited bed capacities to begin with, and in an emergency like this, they were prioritizing private parties over government-subsidized treatments.

While a welcome first step, Ayushman Bharat is clearly far from adequate to cover the various debilitating emergencies faced by the poor. Are there ways in which stronger insurance coverage can be provided? How will we pay for it? We offer a simple, implementable solution. We call it **Ayushman Bharat Plus**.

We have argued that interest on savings is often too small to be attractive for poor people. Our insight is that, however, interest on savings could be sufficient to cover the costs of insurance premiums. The combination of the fact that the poor don't care about the small return they would earn on their savings combined with their reluctance to pay out-of-pocket for insurance coverage, like in the story of Baladev.

Baldev Rai is a blacksmith. He considers himself an honest man who makes a living from moulding utensils by hand, in the traditional way he was taught by his father, who in turn was taught by his. Baldev doesn't know if his family has ever done anything other than this. Since the time he can remember, his existence has been hand-to-mouth. His father could barely make ends meet, and they went without food on more than one occasion. Baldev is grateful that he hasn't faced such dire times in several years. The last time it happened was because he was bedridden with typhoid for a month or so, which meant that they had to use up whatever little he had saved before that. One severe case of typhoid had wiped out his whole life's savings in two months.

This is why he's already teaching his ten-year-old son his craft. He believes it's never too early to learn, and the sooner his son starts to augment their income, the better it will be for all of them.

One of the BCs he sometimes goes to for bank-related work advised Baldev to start saving formally, but Baldev just shook his head at the idea. There was barely any money to begin with. Of what use would it be lying in a bank account far away where he couldn't even see it? At home, even if it didn't multiply, at least he could keep an eye on it. And not like the few hundreds or thousands he could muster would grow much anyway.

The BC reminded him about the time when he lost his livelihood just because he'd fallen ill. At times like these, insurance services could be very useful to Baldev. All he'd have to do is put aside some money for a health emergency, especially given how prone he is to illness because of his vocation. But Baldev just shrugged at the sky and lamented that God would look after him like he had so many before him. And then there were his family and the community. 'At least I won't land up on the streets,' he said philosophically before going right back to his workshop and to his life.

Baldev's thinking that God or family will take care of them in case of emergencies gives us a serendipitous opportunity to combine the two. We suggest two plans.

Basic Plan

Basic Plan			
	Customer A	Customer B	Customer C
Savings amount	**Rs 100,000**	**Rs 50,000**	**Rs 10,000**
Health coverage	**Rs 5,00,000**	**Rs 25,000**	**Rs 50,000**
Real interest rate	3 per cent	3 per cent	3 per cent
Interest amount	INR 3,000	INR 1,500	INR 300
Premium amount for four people	**Rs 15,000**	**Rs 7500**	**Rs 1500**
Personal contribution (interest on savings)	Rs 3000	Rs 1500	Rs 300
Subsidy from government for four people	Rs 12,000	Rs 6000	Rs 1200
Cost per person to government	**Rs 3000**	**Rs 1500**	**Rs 300**

In the basic plan the customer can choose how may they wish to save and will be offered a health insurance coverage of five times of their savings. For example, if a customer saves Rs 1 lakh (approximately $1219) in savings, they will be offered a cover of Rs 5 lakh (approximately $6096) for a family of four. The premium for a Rs 5 lakh cover is Rs 15,000 (approximately $183), of which Rs 3000 (approximately $37) will come from 3 per cent interest (without inflation) on the Rs 1 lakh saving. The remaining Rs 12,000 (approximately $146) will have to be subsidized by the government. That amounts to a subsidy of Rs 3000 for each individual covered. Assuming that fifty crore people need to be covered under this plan, the aggregate yearly cost to the government will be Rs 1,50,000 crore (approximately $28 billion) which is not very much compared to many other subsidies the government spends money on, like the Public Distribution System. The required subsidy can be reduced substantially if, say, the first Rs 10,000 expenditures are required to be a co-pay payment that the family must pay out of their savings.

Enhanced Plan

Enhanced Plan	
	Customer D
Savings amount	Rs 2,00,000
Health coverage	Rs 6,00,000
Real interest rate	3 per cent
Interest amount	Rs 6000
Premium amount for four people	Rs 18,000
Personal contribution (interest on savings)	Rs 6000
Subsidy from government for 4 people	Rs 12,000
Cost per person to government	Rs 3000

Beyond the basic plan, any savings over Rs 1 lakh will get equal additional coverage. So, for example, if your savings are Rs 2 lakh (approximately $2438), your total cover will be Rs 6 lakh (approximately $7315), which is Rs 5 lakh (approximately $6096) from basic coverage (which is subsidized 80 per cent by the government) plus an additional Rs 1 lakh (approximately $1219) coverage, which needs no subsidy because the 3 per cent interest on Rs 1 lakh (approximately $1219) is sufficient to cover the Rs 3000 (approximately $2438) premium for the additional Rs 1 lakh coverage. People who have higher incomes will find this attractive because not only are they getting insurance coverage without paying anything out-of-pocket, but their savings remain intact.

To emphasize, the central idea we are exploiting here is that 3 per cent in interest is too small an amount to induce saving behaviour from the poor but is sufficient to provide critical insurance coverage for which the poor are reluctant to pay out-of-pocket. The added insurance coverage may, in fact, push them to save more in a formal system than they otherwise would without this added incentive.

Chapter Thirteen[1]

Roads and Finance

[1]This chapter is based our op-ed which we had co-authored with Shilpa Kumar and was published in the *Economic Times* on 17 September 2021.

A well-built and well-functioning infrastructure is a necessary condition for economic development in any society. Of course, building and maintaining infrastructure costs money. An important question is: where would the funds for infrastructure come from? A basic principle of economics suggests that a person who uses any *private* good, such as a camera or a bicycle, must pay for it. This willingness to pay generates demand for a private good, which induces others in society to produce that good and supply it using markets. A famous quote from Adam Smith's classic treatise aptly describes this phenomenon.

'It is not from the benevolence of the butcher, the brewer, or the baker that we expect our dinner, but from their regard to their own self-interest. We address ourselves not to their humanity but to their self-love, and never talk to them of our own necessities, but of their advantages.'—**Adam Smith,** *An Inquiry into the Nature and Causes of the Wealth of Nations*, **Volume 1.**

Infrastructure, however, is not a private good. It is a public good that is used and enjoyed by everyone, irrespective of their economics or their politics. Roads are an example of this kind of public good. A driveway leading to your house from the main street may be a private good because only you, your family and any visitors coming to your home benefit from it. So, you pay for it when you build the house, maintain it and have it cleaned regularly. Just like you would do for the rest of your home.

However, what happens when a road that leads from your home to a market is used by many people at the same time? Who will build such a road, and who will pay to build and maintain it? You might be tempted to say that it should be jointly paid for by everyone who uses it, who can jointly delegate the construction of the road to an expert who knows how to build such a road. But 'jointly paid' is hardly a description of any actual financing mechanism that can be easily implemented. It is ambiguous and therefore does not ascribe ownership to anyone in particular. A number of mechanisms have been and continue to be used.

Infrastructure Costs Money. But It Is Money Well Spent

One example is to administer a toll charge every time someone uses a road. Toll roads, in fact, have become increasingly popular over the years. The technology for collecting tolls without having to rely on inefficient and cumbersome human methods involving toll booths has been phased out in favour of automatic detection of usage combined with electronic billing and collection of toll charges.

A toll charge per use is attractive if it can be implemented without friction because it can effectively use differential pricing to combat congestion. It is also fair, as it gets close to pricing the use of infrastructure as if it were a private good. This also minimizes the need for government involvement, which can be riddled with its own inefficiencies and corruption. Ronald Coase, in his influential paper, 'The Lighthouse in Economics',[2] showed how a simple docking fee for all boats effectively financed privately provided lighthouses—commonly thought of as public goods—in seventeenth- to nineteenth-century Britain.

---

[2] Coase, R. H., 'The Lighthouse in Economics', *The Journal of Law & Economics* 17, no. 2 (1974): 357–76, http://www.jstor.org/stable/724895.

In many other situations, however, a per-use charge is inefficient. Think of any well-constructed road at a market that you frequent in your favourite city. It's probably a well-paved road flanked by footpaths and shops on both sides and is used by motor vehicles, bicyclists and pedestrians. How is this road financed? It was most likely built by and is maintained by the local city government. The construction was financed by taxes that are not uniform across all users. Automobiles pay an annual registration fee, part of which has been used in the construction and maintenance of this road as well as many other roads in the city. Bicyclists and pedestrians, on the other hand, probably pay no corresponding fee. In effect, automobile owners are probably subsidizing the use of this and many other roads for bicyclists and pedestrians.

Is this differential taxation optimal? The answer is yes, and for two reasons. To begin with, pedestrians and bicyclists don't really need the kind of high-quality roads that automobile owners require. But once a high-quality road is built to service the needs of automobiles, the *marginal* cost of allowing pedestrians to use the same roads is trivial in comparison to banning the pedestrians from using the high-quality roads and building separate lower-quality roads for them.

Second, this particular infrastructure is more useful for *everyone* together. Only when *everyone* is allowed to use it together, which is necessary for preserving the essence of a marketplace, making it less of what is called a 'club good', where usage is restricted to that of a public good. The simultaneous presence of various different people complements the enjoyment and use of such infrastructure.

Many western tourists visiting the city of Mumbai find it incredible that posh, high-rise buildings and slum dwellings exist in close proximity. If one reflects on it, however, it would explain why this makes sense. The rich and privileged greatly benefit from the services provided by slum dwellers who work as domestic helpers, car cleaners, vendors who sell vegetables and essentials, autorickshaw drivers and semi-skilled workers such as plumbers and electricians. The service-providing group, in turn, depends on the economic benefits they receive from the rich and upper middle-class inhabitants of the high rises. This co-dependence requires infrastructure that makes sense only when it is available to and shared by all.

Finance Is Infrastructure

Financial infrastructure is like roads. To serve high- and middle-income clients, banks need physical branches, ATMs, smartphone apps and internet banking services. These cost money, and the fees collected by banks from customers, either direct fees or indirect fees in the form of a spread between borrowing and lending rates, finance the development and maintenance of such infrastructure. The average cost of servicing a simple bank account is about Rs 500 (approximately $6) per year, based on research and interviews we conducted with bank officials at the Digital Identity Research Initiative (DIRI) at the Indian School of Business. A simple, basic account, such as a PMJDY account, does not generate revenue in fees and spreads that are even close to Rs 500 per year. But because initiatives like the PMJDY are government-run, banks are often forced by regulators to serve BOP customers. The banks, in turn, do this reluctantly because they make no money from these customers but instead incur costs.

We argue that Rs 500 (approximately $6) is the wrong benchmark to overcome. The *marginal* incremental cost to also serve an additional basic account customer, with the existing infrastructure already built for regular customers, is only around Rs 120 (approximately $1.5) per year or Rs 10 (approximately $0.12) per month. Even that may not be low enough for any profit-making private financial institution to actively seek out poor customers. But it is low enough for the government, i.e., the taxpayer, to bear this cost to accelerate financial inclusion.

This is exactly like not charging a road tax to pedestrians in our previous example. But the use of the same roads (most of them, in any case, except perhaps for highways) by pedestrians as well as automobile owners creates positive externalities for both parties involved—the ones paying and the ones not.

For instance, a car owner can drive their car on the road as well as walk on the footpath next to the same road when they desire. In the case of financial and banking services, it would be efficient for you to pay your maid, your driver and your *sabjiwala* (vegetable seller) using digital mobile money if they could also use the same digital mobile money to buy goods and services they need and carry out simple financial transactions such as transfers without paying *any* transaction costs.

In other words, if the financially included could subsidize the use of digital products for the financially marginalized or excluded group (thereby including them), it would be in their own best interest. Recall Adam Smith's words of wisdom when we started this chapter with, 'It is not from the benevolence of the butcher, the brewer, or the baker that we expect our dinner, but from their regard to their own self-interest . . .'

The inclusion of marginalized groups into the financial system need not be regarded as charity at all because, in fact, it isn't. It's in everyone's best interest. Even that of the already well-included.

So, we are proposing *zero* transaction costs for BOP customers. Some of you may not be convinced that zero costs are justified and may argue for making everyone pay at least the marginal cost associated with each electronic transaction. But what exactly are the marginal costs of an electronic transaction? The digital payments infrastructure requires high fixed costs in technology, and it may appear that the marginal costs are close to zero. They are not. The reason is that even though the marginal information transmission costs may be close to zero, the costs associated with authentication and cyber security are not zero. That means that even if we levy a very small fee, say one rupee, for a small transaction of a Rs-20 (approximately $0.24) deposit to an account, the account owner may not be okay with it. The reason is that good old cash works for him. Cash is convenient. Cash is private. Cash is intuitive. Cash does not incur explicit transaction costs. So, why would he be willing to pay even one rupee for a small transaction? Eighty-five per cent of all transactions globally (and 40 per cent in the US) are still carried out using cash, particularly transactions involving small amounts of money. There are good reasons why that is the case.[3]

The argument that cash is cumbersome to carry and store, can be stolen or forged, remains uninvested, usually loses purchasing power over time, and, most importantly, cannot be transferred easily across large distances, has no bite for a large section of lower-income people. Digital financial services are intimidating for many poor people. They do not find these simple. They also mistrust technology. We have illustrated through the course of this book how the human touch can both alleviate fears and generate trust in technology.

Kalavati, who cooks for one of us, wanted her salary in cash. When asked why she would not prefer to have her salary

---

[3] Bhagwan Chowdhry, 'We Still Don't Have Safe and Reliable Money', *Time*, July 1, 2015, https://time.com/3942005/bitcoin-not-future-digital-currency/, accessed on July 20, 2023.

directly deposited into her bank account, she provided two compelling reasons. One, she said it was not easy for her to check quickly whether the money did indeed reach her bank account. Two, she confessed that she wanted to use her salary right away to buy food and other essential items for her household, and she said that merchants where she shopped, sabjiwalas, etc., preferred cash. This points to another important requirement for digital payments: they must also become ubiquitous.

An important insight came to us serendipitously when, one day, Kalavati suddenly said that she would be willing to accept her salary digitally if we could transfer it to her young son, who uses Paytm and GPay. She explained that her son would immediately call her when he received a message on his smartphone that the money had been transferred. A trusted human touch and an immediate confirmation alleviated two of her major misgivings about accepting her salary digitally. We also learned that her mistrust was not with us but with the digital process.

During our research, we've gathered that the use of BC agents in kiranas, pharmacies, shopping malls, post offices, bus and train stations, and other busy spaces could help people carry out simple transactions using mobile phones or point-of-sale devices. This could go a long way towards making everyone comfortable with the use of digital financial services. As in the case of Kalavati who warmed up to the idea when her son began using Paytm and Gpay. In other words, had there been human touch through a BC *before* her son became savvy, Kalavati, too, could've boarded the tech train much earlier.

Yes, this human touch will cost money, but a lot less than providing this service in a bank branch using a dedicated teller or even an ATM. DIRIs research estimates suggest that it would cost only around Rs 4800 crore (approximately $585 million) each year to service forty crore (400 million) customers with basic accounts. That amounts to Rs 5000 (approximately $61) per month for each of the eight lakh (0.8 million) BCs that we currently have. In reality, we need far more BCs than we have, perhaps five to ten times as many, each doing digital transactions for others not as a full-time activity but as part-time work. A large group of people who have the bandwidth, the ability and the desire to supplement their income (for example, kirana owners, chaiwalahs, postal workers and Digital Didis, who we've spoken about in detail earlier) can act as BCs so that the presence of such agents becomes ubiquitous and highly visible. To ensure that incentives for BCs are sufficiently attractive, they should be encouraged to teach people how to perform digital financial transactions on their own, following which the BC would get paid for every digital transaction performed by a person that she or he teaches for a period of, say, one to two years.

Government Spending Can Be Wasteful, but in This Case, It Is Just the Opposite

The important point we are making is that the government should bear the cost of transactions associated with basic accounts. One way of doing this may be for the government to compensate BCs and last-mile financial service providers for delivering government payments and subsidies to the poor. A small proportion of the overall Direct Benefit Transfers and other government subsidies directed towards this payout would energize the already existing infrastructure. This would probably be the cheapest way for governments to ensure that the payments reach the vulnerable because the costs are only marginal and small. The customers would then pay *zero* transaction costs.

This is like installing a pedestrian crossing at a marginal cost on a busy road—a small price to pay for ensuring the common man can also be a part of the well-built road infrastructure. The most important governmental expenditure is on infrastructure with large positive societal benefits—economists call them externalities. A marginal addition of infrastructure that ensures smooth financial services for everyone is one such expenditure where returns are enormous, as the creation of NPCI and the meteoric growth of UPI-based transactions have already demonstrated.

Technology has arrived. Let us make it SHUb. Not just for the middle- and upper-income populations but for everyone.

Acknowledgements

With profound gratitude, we acknowledge the invaluable contributions of those who have supported us throughout this transformative journey in creating this book.

First and foremost, we would like to express our deep gratitude to our esteemed colleagues, Prof. Shilpa Aggarwal, Prof. Shashwat Alok, Prof. Krishnamurthy Subramanian and Aashita Joshi of the Indian School of Business. Their invaluable insights, support and constructive criticism have played a crucial role in refining the ideas and solutions developed during our work at the Digital Identity Research Initiative. These contributions have now been presented in this book.

We extend our sincere thanks to Prof. Amit Goyal, Amit Bubna, Shilpa Kumar and Lisa Nestor, who co-authored some of our past publications. Their expertise has been foundational in the development of the concepts we have referred to and expanded upon in this book.

We would also like to express our gratitude to **everyone** who generously provided their time and expertise. Additionally, our heartfelt appreciation goes out to the organizations that collaborated with us in piloting some of our recommendations and granting us access to crucial data and resources. Their support has been invaluable, ensuring the depth and accuracy of our research.

A special mention of thanks is to the team at Penguin Random House for their unwavering support and guidance throughout this incredible journey. We are particularly grateful to the editors, Tarini Uppal and Archana Nathan, as well as our copy editor, Yash Daiv, for their unwavering support and patience.

Last but not least, we want to express our appreciation to all the readers of this book. Your interest in our work fuels our passion for the subject matter, and we sincerely hope that this book enriches your understanding. You can scan the QR code on the cover of this book or visit www.FintechforBillions.com to know more. We eagerly await your views and feedback. Feel free to reach out to us via email: Authors@FintechForBillions.com or tweet to us on Twitter: @BhagwanUCLA & @AnasAhmed_S